Lecture Notes in Networks and Systems 1092

The series “Lecture Notes in Networks and Systems” publishes the latest developments in Networks and Systems—quickly, informally and with high quality. Original research reported in proceedings and post-proceedings represents the core of LNNS.

Volumes published in LNNS embrace all aspects and subfields of, as well as new challenges in, Networks and Systems.

The series contains proceedings and edited volumes in systems and networks, spanning the areas of Cyber-Physical Systems, Autonomous Systems, Sensor Networks, Control Systems, Energy Systems, Automotive Systems, Biological Systems, Vehicular Networking and Connected Vehicles, Aerospace Systems, Automation, Manufacturing, Smart Grids, Nonlinear Systems, Power Systems, Robotics, Social Systems, Economic Systems and other. Of particular value to both the contributors and the readership are the short publication timeframe and the world-wide distribution and exposure which enable both a wide and rapid dissemination of research output.

The series covers the theory, applications, and perspectives on the state of the art and future developments relevant to systems and networks, decision making, control, complex processes and related areas, as embedded in the fields of interdisciplinary and applied sciences, engineering, computer science, physics, economics, social, and life sciences, as well as the paradigms and methodologies behind them.

Indexed by SCOPUS, EI Compendex, INSPEC, WTI Frankfurt eG, zbMATH, SCImago.

All books published in the series are submitted for consideration in Web of Science.

For proposals from Asia please contact Aninda Bose (aninda.bose@springer.com).

Ruslan Polyakov
Editor

Ecosystems Without Borders 2024

Opportunities and Challenges

Springer

Editor
Ruslan Polyakov
Kaliningrad State Technical University
Kaliningrad, Russia

ISSN 2367-3370 ISSN 2367-3389 (electronic)
Lecture Notes in Networks and Systems
ISBN 978-3-031-67353-5 ISBN 978-3-031-67354-2 (eBook)
https://doi.org/10.1007/978-3-031-67354-2

This Springer imprint is published by the registered company Springer Nature Switzerland AG
The registered company address is: Gewerbestrasse 11, 6330 Cham, Switzerland

Preface

Ecosystem development is a complex, multi-dimensional process that requires the interaction of different scientific disciplines as well as cooperation between different countries and cultures. «Ecosystems Without Borders 2024 - Opportunities and Challenges» in your hands is a collection of papers delivered at the Third International Conference "Ecosystems Without Borders: Opportunities and Challenges" held at Kaliningrad State Technical University in February 2024.

The materials include studies reflecting the transformation of ecosystems, both in innovative development and in science, technology, and business, as well as the spatial aspects of ecosystems and features of the formation of a creative class in ecosystem conditions. In addition, the book includes theoretical articles that aim to implement the ideas of sustainable development and circular economy.

The book is intended for a wide range of readers, including practicing economists, students, graduate students, and researchers, as well as government officials and company managers. It can serve as an important source of information and knowledge that will lead to scientific and innovative advances in the sustainable development of society.

We thank everyone who contributed to this book, including paper authors, session presenters, section chairs, Program Committee members, volunteers, as well as reviewers and sponsors. Special thanks go to the conference organizers, without whose help this book would not have been possible.

We hope that this book will serve as an important tool for advancing the science and practice of ecosystems and sustainable development, and will benefit readers.

April 2024 — Ruslan Polyakov

Contents

Spatial Ecosystems

Innovation Ecosystems

Theoretical Aspects of Ecosystems

Methodological Aspects of Formation of Regional Policy of the State from the Position of Eco-System Approach (on the Example of Agricultural Sector of Kaliningrad Region)

Inobbat Alieva[1](✉), Dmitry Kuznetsov[2], Elena Skorobogatykh[2], and Anton Martynov[1]

[1] Immanuel Kant Baltic Federal University, Kaliningrad, Russia
inobbat@mail.ru
[2] Kaliningrad State Technical University, Kaliningrad, Russia

Abstract. The theory of ecosystems is currently becoming increasingly relevant and is recognized by the scientific community as a rather promising direction for the development of both theory and practice. In this paper an attempt is made to study the principles of formation and implementation of regional policy from the position of ecosystem approach and to use the tool of analyzing the state of the agricultural sector of the Kaliningrad region in the context of economic efficiency indicator and evaluation of the implemented regional policy. A substantive theoretical and methodological review of all possible sides and aspects of regional policy in the context of its structural components, such as socio-economic, ecological and resource-saving. At the level of implementation of public policies, the need for both preliminary and final evaluation of each stage of implementation according to the principle of synergistic effect and subsequent diffusion to other types and measures of public policy becomes more obvious. Moreover, the digital platform and tools of remote analysis in real time, which act as a key parameter of eco-systemic approach and acquire serious prospects, make sufficient adjustments not only in the mechanism of regional policy implementation, but also in the process of its development at the level of individual evaluation procedures. This paper attempts to systematize the mechanism of regional policy development, to build a mediated hierarchy of time and priority of its components, as well as to clarify the platform structure as a tool for operational communication and instant assessment of indicative parameters of implementation. As the object of the study, the agricultural production and processing industry is presented as the one that best meets the requirements of eco-systematicity: economic, social, environmental and managerial-resource-saving.

Keywords: Eco-System Approach · Regional Economy · Regional Policy · Platform Solutions · Regional Policy Elements · Agriculture · Agrarian Policy · Network Effects

R. Polyakov (Ed.): EcoSyst 2023, LNNS 1092, pp. 3–13, 2024.
https://doi.org/10.1007/978-3-031-67354-2_1

1 Introduction

The importance of effective implementation of regional policy in complex conditions of high turbulence of the economic environment and economic practice is becoming more and more relevant every day. In this regard, there is an objective need to expand the range of scientific and methodological approaches to its assessment both from the position of development and from the position of implementation efficiency. In addition, given the expanded technical capabilities, including through platform solutions, a methodology is required that allows for a comprehensive design of regional policy almost in parallel at its input and output, in the form of specific measures and actions that allow for a timely and systematic response to all possible deviations in the course of its implementation. As is known, ecosystems represent actors of economic reality, which on the basis of platform solutions are comprehensively organized with all economic entities, whose economic interest directly or indirectly coincides with the main participant of this system of relationships. An important distinguishing characteristic of the ecosystem is the permanent correlation and adjustment of processes in the direction of achieving a common result and, as a consequence, the coordination of all actions. If we focus all attention on regional policy and try to apply an ecosystem approach to it, it means that this set of measures and activities, which in a general sense is called regional policy, should be studied as a set of forces, actors, elements that are involved in the overall scheme of its implementation through the solution of social, economic, environmental and management-saving tasks that need to be organized in such a way that prioritization is achieved through the achievement of the principle of P, which is the principle of "ecosystems". Within the framework of this work it is supposed to focus on the implementation of public policy in the field of agriculture, since this industry as no other meets, in our opinion, the principles of eco-systemic approach.

2 Materials and Methods

Emphasizing attention to the level of the region in the formation of economic development policy, more than relevant discourse in modern conditions. The socio-economic state of the region predetermines the state of the national economy as a whole. Moreover, the meso-level of the national economy is an integrator of all managerial decisions on the subject of their economic efficiency and rational expediency, as it is in the key segment of execution and determination of long-term sustainability of the economic system as a whole. The priority of the application of eco-systemic approach in the formation of regional economic policy is primarily due to its specificity, which relies on the complexity and systemic nature at the same time, in other words, its synergetic effect is an important criterion of its universality [1]. High turbulence of both the external environment and, as a consequence, reactionary changes in the internal environment make us reconsider not only the approaches, but also the tools and mechanism of realization of regional policy in general. Large-scale digital and structural changes in one way or another lead to changes in institutions, and those, in turn, to cultural transformations and challenges, to which it is necessary to respond with flexible solutions, in a timely manner and with the participation of as many stakeholders as possible in the process of forming a new value as a management decision at the state level [2].

The ecosystem approach in the context of regional economic discourse considers the regional system as "a dynamically developing socio-economic community consisting of actors (players) operating in its different sectors, which jointly develop competencies, while working on a cooperative and competitive basis. This allows to carry out structural, technological, institutional and other transformations of the economy without disturbing the socio-ecological-economic balance, developing culture and improving the quality of life of the population" [3].

The application of the ecosystem approach to the regional economy requires researchers to focus on several key aspects (structural components) [4]. First, on the priorities of regional management (final and intermediate goals), which may include:

- ensuring economic growth;
- achieving the Sustainable Development Goals;
- human capital development;
- increase of investment and innovation potential, etc.

Second, on the subjects of the regional economic ecosystem (actors, stakeholders), which can be:

- population;
- business community;
- public authorities;
- educational, scientific organizations and other subjects of innovation infrastructure;
- public organizations, etc.

Third, on the objects of the regional economic ecosystem, which can be:

- national projects in which the region takes an active part;
- regional development strategies and programs;
- sectoral development strategies and programs in which the region takes an active part, etc. [5].

Fourth, on technologies, methods and tools of interaction of actors (stakeholders) within the regional economic ecosystem, including:

- technologies and tools for involving all stakeholders in decision-making processes;
- The use of digital platforms;
- sharing of infrastructure facilities and their material and technical base;
- project and program-targeted management;
- application of special financial instruments ("green", "social", "sustainable" investment, targeted financing, preferential lending and taxation), etc. [6].

Fifth, on the functions of key actors (stakeholders) related to:

- socio-eco-economically responsible management, production, investment, consumption;
- fulfillment of social obligations;
- developing internal competencies;
- by conducting R&D;
- improvement of the regulatory and legal framework, development of institutional and competitive environment, etc. [7, 8].

The application of the ecosystem approach methodology will make it possible to comprehensively solve research and management tasks, relying on positive synergy effects, increase the efficiency of limited resources utilization, and, based on negative system effects, identify areas of compromise solutions to increase stakeholder satisfaction, ensure social harmony, increase the resilience of economic systems and their long-term sustainable development [9].

Table 1. Indices of physical volume of gross regional product by types of economic activity (in constant prices; in % to the previous year)

	2020	2021	2022
Gross regional product	101.6	100.2	109.0
Agriculture, forestry, hunting, fishing and fish farming	105.5	114.2	98.9
Extraction of minerals	94.5	92.0	101.5
Manufacturing industries	101.0	93.6	100.8
Provision of electricity, gas and steam; air conditioning	96.8	95.0	110.1
Water supply; wastewater disposal, organization of waste collection and utilization	104.7	98.7	114.5
Construction	77.3	98.6	97.9
Wholesale and retail trade; repair of motor vehicles and motorcycles	91.7	112.5	106.0
Transportation and storage	118.4	92.8	126.0
Activity of hotels and catering enterprises	104.7	74.7	131.7
Information and communication activities	134.2	110.9	107.4
Financial and insurance activities	82.4	128.4	105.5
Activity on operations with real estate	111.8	101.9	107.8
Professional, scientific and technical activities	107.4	105.1	113.3

Executed by the authors on the basis of [4].

As presented in Table 1, the dynamics of participation of agriculture in the gross regional product of the Kaliningrad Oblast is ambiguous. But one way or another agriculture plays an important role in the formation of the gross regional product of the Kaliningrad region.

At the same time, the state of agriculture has been changing quite seriously in recent years. In particular, in 2022, sown areas of grain and leguminous crops in the Kaliningrad region decreased by 8.8%, amounting to 136 thousand hectares, reported Kaliningradstat. A year earlier this indicator was equal to 149.2 thousand hectares [10].

13 years ago, 64 thousand hectares were planted with cereals and beans in the region: 2.1 times less than last year. In 1980, 149.6 thousand hectares of fields were sown with such crops, and in 1970 - 153.3 thousand hectares.

Sown areas of oilseed crops for the year increased by 33.3% - from 49.2 to 65.6 thousand hectares. In 2010, the plants occupied 29.7 thousand hectares. This is 2.2

times less than in the previous year. In 1980, oilseeds accounted for only 0.1 thousand hectares of land.

The area under potato crops last year increased by almost 3.5%, amounting to 6 thousand hectares (in 2021 - 5.8 thousand hectares). In 2010, this figure was higher by 27.7% and reached 8.3 thousand hectares. At the same time, in 1970 as much as 22.9 thousand hectares were occupied by potatoes.

In 2022, 2.8 thousand hectares of vegetables and melons were sown in the region - 3.7% more than a year earlier (2.7 thousand hectares). 13 years ago, the figure was 12% lower - 2.5 thousand hectares. In 1970, vegetable and melon crops occupied 4 thousand hectares of land [10].

Gross harvest of grain and leguminous crops in 2022 fell by 3.5% to 688.8 thousand tons (in 2021 - 714.2 thousand tons). In 2010, the crop was 3.7 times smaller and amounted to only 186.4 thousand tons. In 1980, the region harvested 208 thousand tons of grains and beans, and in 1970 - 253.3 thousand tons.

The yield of oilseeds for the year increased by 18.1% - from 159.9 to 189 thousand tons.

Gross harvest of potatoes last year also increased by 4.8%, amounting to 118.8 thousand tons (in 2021 - 113.3 thousand tons). However, 13 years ago, the yield of this tuber crops was 17% higher - 143.3 thousand tons, and in 1970 it reached 299.1 thousand tons.

The yield of melon vegetables in the Kaliningrad Oblast grew insignificantly over the year - by 0.6% to 68 thousand tons (in 2021 - 67.6 thousand tons). In 2010 their gross yield was lower by another 1.3% and amounted to 67.1 thousand tons. In 1980 in the region harvested 52 thousand tons of vegetables and melons, in 1970 - 58.8 thousand tons.

As a result, we can confidently say that the agricultural sector of the Kaliningrad region's economy is a rather important element of the aggregate production system.

Today, the share of agriculture is about 6.6% of total production, which indicates its importance both in terms of production and formation of income of the population. Here it should be noted that the main role in the production of agricultural products falls on large agricultural holdings.

As defined above, the agricultural sector is a rather demonstrative object as an evidence base for the need to apply an ecosystem approach to regional policy making. First of all, we should mention its economic importance as a sector that supplies life-supporting resources and, moreover, is a sufficiently large reservoir for job creation. Food security and the use of labor force, including low-skilled labor, are important parameters of this industry. And as of January 1, 2023, of the total number of organizations 33625 units, 1950 belong to the agriculture sector. In particular, if to specify and refer to the state of the agricultural sector in Kaliningrad region, we can get the following information.

According to the presented statistical data, there is a positive dynamic in the change of the volume of agricultural production, which indicates, among other things, the growth of demand for the products of the industry. As shown in Fig. 1, the structure of agricultural production is predominantly in favor of crop products, which means that land quality is a significant factor of growth.

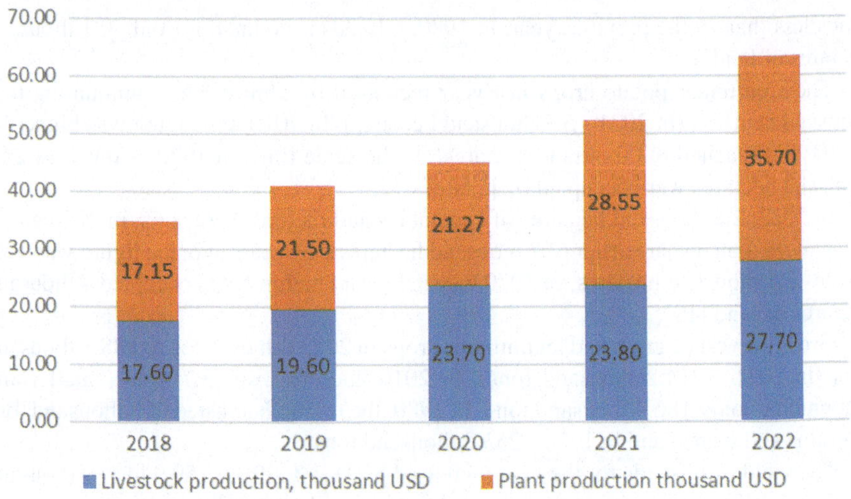

Fig. 1. Agricultural output (farms of all categories)

Table 2. Growth rate of the agricultural market

	2020	2021	2022
Index of agricultural production	109.2%	99.7%	101.9%
Average annual growth rate		−4.4	3.6

Executed by the authors on the basis of [4].

Having calculated the average annual growth rate of the agricultural production market, we will try to assume that in the future this industry will only expand and increase the volume of its production (see Table 2). This is also explained by the fact that the exclave character of the region implies orientation first of all on satisfaction of the internal demand and only after that on the market of big Russia.

At the same time, we can also talk about the strengthening of market concentration and consolidation of production (see Table 3).

Table 3. Number of organizations in the agricultural sector

	2020	2021	2022
Number of organizations in the industry	1031	892	827

Executed by the authors on the basis of [4].

Thus, the formation of regional economic policy in the field, including support of agriculture in Kaliningrad region, should be based on the principle of satisfying a greater number of stakeholders - interested players. In particular, the growth of the market and strengthening of market concentration indicate the satisfaction of interests of large producers first of all, but not of small and medium-sized businesses [10]. Moreover, the change in the profitability of production also indicates insufficient attention in the framework of regional policy to the issues of support of economic efficiency of small agricultural market entities. (See Table 4.) And here it is important to note that the support mechanism is not through direct financing or subsidies, but support through the introduction of high technologies to improve the quality of products.

Table 4. Average sectoral profitability of agriculture

	2020	2021	2022
Producer price indices for agricultural products	110.7%	110.6%	113.8%
Index of agricultural production	109.2%	99.7%	101.9%
Industry average profitability	6.2%	6.8%	6.6%

Executed by the authors on the basis of [4].

Table 5. Key indicators for analyzing the industry

	2020	2021	2022
Average industry profitability	16.3%	11.4%	13%
Industry share in the total GRP	6.2%	6.8%	6.6%
Volume of own production, mln. Rub	46 134.8	56 410.8	58 535.8
Average number of employees, thous. People	21538	21555	22086
Number of organizations in the industry	1031	892	827
Index of agricultural production	109.2%	99.7%	101.9%

Executed by the authors on the basis of [4].

Profitability is an important indicator of efficiency and competitiveness of a given sector of the economy. One of the important factors affecting the profitability of virtually any sector of the economy is market conditions. The level of demand for the products of enterprises in this sector depends on the general economic situation, the level of income of the population, investment activity and other factors. A steady increase in demand for the offered products may have a positive impact on profitability, while a decline in demand - vice versa. Therefore, special attention should be paid to analyzing the economic conjuncture and forecasting its changes.

In this regard, we would like to pay special attention to the financial support from the state and try to assess its economic efficiency.

In particular, on the basis of statistical data from the information handbook on measures and directions of state support of the agro-industrial complex of the Russian Federation [10], a paired linear regression equation was constructed (Fig. 2) and it was found that in Russia as a whole the change in the volume of production by 49.4% is determined by the volume of state support ($R^2 = 0.494$). This is quite a high indicator.

Table 6. Initial data on subsidies and volume of production from 2015 to 2022, thousand.

Year	Subsidies for agriculture (x)	Volume of production (y)
2015	1353 584	20 814
2016	2210 981	24 402
2017	2141 528	31 048
2018	2233 900	28 580
2019	2614 878	32 396
2020	1971 313	34 739
2021	1916 340	40 752
2022	2510 823	46 135

Executed by the authors on the basis of [4]

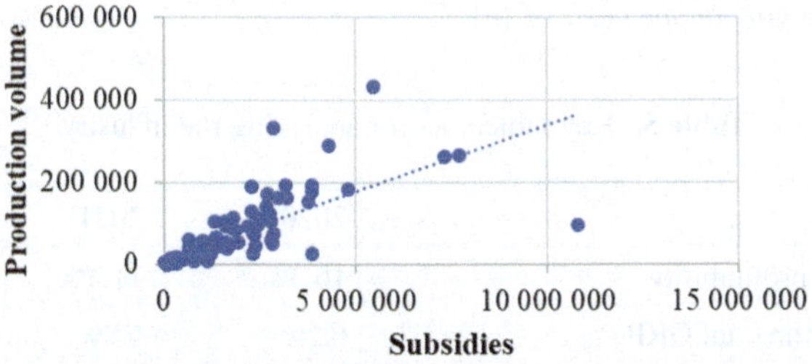

Fig. 2. Linear paired regression plot between production volume and subsidies in Russia

Does the Kaliningrad region correspond to the all-Russian indicators? To answer this question, the necessary statistical data on the Kaliningrad region were collected, regression models were built and analyzed.

When constructing the paired regression model, the volume of agricultural output was chosen as the resultant feature (y), and the amount of state subsidies was chosen as the factor feature (x). Based on statistical data for the Kaliningrad Oblast for the period from 2015 to 2022 [10], a paired linear regression equation was constructed for these indicators (Fig. 3).

According to the results of calculations it was revealed that there is a direct moderate dependence between these indicators, the growth of production depends on subsidies for agriculture only by 23.7%. This is significantly lower than the overall indicator for the country.

Thus, a direct increase in financing of the agricultural sector does not give an economically effective result.

No less important is the issue of employment in the agricultural sector. Thus, according to G.M. Fedorov "…the reduction in the number of people employed in agriculture is connected with the concentration of production at larger producers - agricultural enterprises, more mechanized and not requiring such a large number of employees as

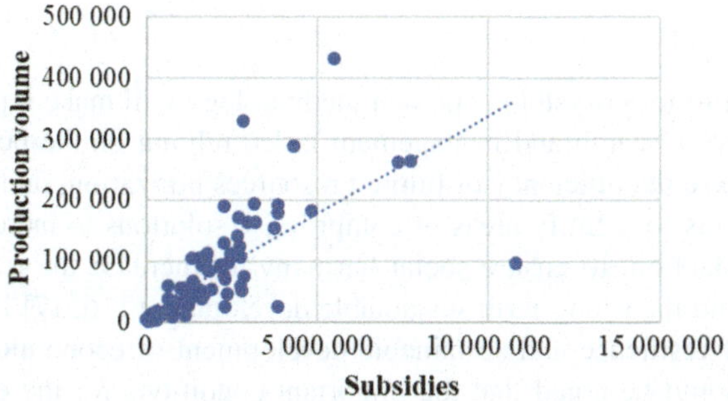

Fig. 3. Linear pairwise regression plot between production volume and subsidies in Kaliningrad region

compared to peasant (farm) farms and household farms" [11]. This process in the Kaliningrad region is more intensive in comparison with the RF as a whole, where it also takes place (Table 3). At the same time, the growth rates of agricultural production in the Kaliningrad Oblast due to the development of large enterprises are higher than the national average: if in 2005 the Oblast accounted for 0.54% of agricultural output, in 2021 - 0.78%

All of the above predetermines the need for a more comprehensive or ecosystem approach to the formation of regional agricultural policy. As defined above, the ecosystem approach primarily implies the organization of management of the object in an integrated manner through the use of all available resources, which implies their maximum conservation and sustainable use on a rational basis, which means that the tools of regional policy in the field of agriculture should also become measures to improve not only quantitative parameters and results, but also qualitative improvement through the solution of such problems as the level of staffing of the industry, the use of modern resource-saving technologies, a well-established marketing system oriented, among other things, to the integrated processing of agricultural raw materials and products [12, 13].

Ecosystem approach to agricultural policy making, which implies taking measures based on the principle of universal "tolerance" and taking into account the interests of all stakeholders and actors. In particular, by letter designations we can take the following tasks and problems, and as a consequence the corresponding measures: A- training of qualified personnel in the agricultural sector, B increasing the profitability of agricultural enterprises, C increasing the wages of agricultural workers, D- increasing the level of applied technologies (accessibility), E- accessibility and quality of agricultural products, F- research and development in the field of ecological products and processing industry, G- regular, uninterrupted and sustainable development of agricultural products, F- sustainable agricultural production and processing industry, G- sustainable agriculture [14]. The development of a solution mechanism by them should "lie in the green zone" of interests, not prevailing in any of the parties both from the position of resource consumption and prioritization [15].

Accordingly, in this regard, regional policy should be considered from the position of the most resource-saving instrument of state regulation, which will be focused on the effective solution of regional problems of our time.

3 Results

The application of the ecosystem approach methodology will make it possible to comprehensively solve research and management tasks, relying on the positive effects of synergy, to increase the efficiency of limited resources utilization, on the basis of negative system effects to identify areas of compromise solutions to increase the level of stakeholder satisfaction, to ensure social harmony, to increase the resilience of economic systems and their long-term sustainable development [16, 17] Diversity plays a major role in the resilience and sustainable development of economic ecosystems. In this regard, it should be noted that the important conditions for the construction and successful functioning of entrepreneurial, innovative and industrial ecosystems on the scale of the regional economy are a sufficient concentration of scientists, researchers, engineers, expanded reproduction of these personnel due to the increasing number of graduates of higher education and postgraduate programs, non-decreasing investments in research and development, the development of innovative, industrial, transport and entrepreneurial infrastructures, as well as the development of innovation, industrial, transport and entrepreneurial infrastructure In addition, the development and continuous improvement of institutions and institutional environment in which the economic ecosystem functions plays a major role [18]. All this together will ensure the growth of innovation-technological and entrepreneurial potential of the region and create the necessary conditions for productive ecosystem interaction of scientific organizations, industrial enterprises, entrepreneurs and investors.

4 Results and Discussion

Thus, regional economic policy on the basis of eco-system approach is a set of measures focused on the rational use of space and self-development of the region, which implies the formation of such a mechanism of state regulation, which will be focused primarily on improving the quality of life of the population of the region through rational consumption and cultural and spiritual improvement, ecological production and technological development. The development of ecosystems is a logically justified trend of the digital economy, which makes it possible to build a management structure based on big data and operational information exchange on the basis of synergistic effects, which allows it to increase competitiveness by meeting the needs of customers and by expanding the perimeter of coverage of the structural elements of the ecosystem.

The application of the ecosystem approach to regional policy making is more than justified and scientifically argued, as the ecosystem approach is an approach based on the principle of "tolerance" and harmonious correlation and balance of interests of a sufficient number of stakeholders. The ability to satisfy these interests in terms of both temporal and quantitative-qualitative parameters is determined precisely by ecosystem decisions in the course of implementing the regional policy of the state.

References

1. Tretyakova, E.A., Freiman, E.N.: Ecosystem approach in modern economic research. Management 1, 6–20 (2022)

2. Simchenko N.A., Anisimova N.Y.: Ecosystem of digital production-labor relations: theoretical and methodological justification. Theor. Pract. Soc. Dev. **5**(159), 61–66 (2021). https://doi.org/10.24158/tipor.2024.1.5
3. Salamatov, A., Maltsev, Y., Pavlov N.: Region innovative development in the Russian economy technological transformation: ecosystem approach. In: E3S Web of Conferences "Ural Environmental Science Forum "Sustainable Development of Industrial Region". pp. 1–8 (2021). https://www.e3s-conferences.org/articles/e3sconf/pdf/2021/34/e3sconf_uesf2021_12004.pdf, https://doi.org/10.1051/e3sconf/20212581204
4. Federal State Statistics Service, - Kaliningrad region in figures, 2023/- Federal State Statistics Service. - Kaliningrad: territorial body of the federal service of state statistics for the Kaliningrad region, pp. 150 (2023)
5. Phills Jr. J.A., Deiglmeier K., Miller D.T.: Rediscovering social innovation. Stanford Soc. Innovation Rev. **6**(4), 34–43 (2008). https://doi.org/10.48558/GBJY-GJ47
6. Lang, T., Görmar, F.: Regional and Local Development in Times of Polarisation. Re-thinking Spatial Policies in Europe, pp. 382 (2019). https://doi.org/10.1007/978-981-13-1190-1
7. Klimuk, V.V.: Application of ecosystem approach in building a model of neo-industrial development of industrial enterprises. Nat.-Humanitarian Res. **37**(5), 136–141 (2021)
8. Ushvitsky, L.I., Ter-Grigoryants, A.A., Denyshchik, M.N.: Formation of the conceptual basis of the ecosystem approach to the development of socio-economic systems. Bull. North Caucasus Federal Univ. **3**(84), 142–154 (2021). https://doi.org/10.37493/2307-907X.2024.2.18
9. Kleiner, G.B., Rybachuk, M.A., Karpinskaya, V.A.: Development of ecosystems in the financial sector of Russia. Upravlenets No **4**(11), 2–15 (2021). https://doi.org/10.29141/2218-5003-2020-11-4-1
10. Kaliningrad region in figures. 2023: Brief Statistical Collection. Kaliningradstat - Kaliningrad, pp. 138 (2023)
11. Fedorov, G.M., Kinder, S., Kuznetsova, T.: On the role of geographical location and employment changes in the dynamics of rural settlement. Baltic Region, T. **13**(4), 129–146 (2021)
12. Acs, Z.J., Stam, E., Audretsch, D.B., O'Connor, A.: The lineages of the entrepreneurial ecosystem approach. Small Bus. Econ. **49**(1), 1 (2017). https://doi.org/10.1007/s11187-017-9864-8
13. Adner, R., Kapoor, R.: Value creation in innovation ecosystems: How the structure of technological interdependence affects firm performance in new technology generations. Strateg. Manag. J. **31**(3), 306–333 (2010). https://doi.org/10.1002/smj.821
14. Abalikstiene, E., Perkumiene, D.: Challenges and problems of agricultural land use changes in Lithuania according to territorial planning documents: Case of Vilnius district municipality. Land Use Policy **117**, 106125 (2022). https://doi.org/10.1016/j.landusepol.2023.106125
15. Aleknavičius, A., Aleknavičius, M.: Problems of rural areas management in Lithuania. Baltic Surveying **2**, 6–14 (2016)
16. Carson, A.D., Carson, D.B., Argent, N.: Cities, hinterlands and disconnected Urbanrural development: Perspectives from sparsely populated. J. Rural. Stud. **93**, 104–111 (2022). https://doi.org/10.1016/j.jrurstud.2024.01.012
17. Daunora, Z.J.: Policentriška urbanistinė struktūra - strateginis lietuvos ire s prioitetas, Urbanistika irarchitektūra, **31**(4), 178—191 (2007)https://doi.org/10.3846/13921630.2007.10697105
18. Glückler, J., Shearmur, R., Martinus, K.: Liability or opportunity? Reconceptualizing the periphery and its role in innovation. J. Econ. Geogr. **23**(1), 231–249 (2023). https://doi.org/10.1093/jeg/lbac028

Factor Characteristics of Neo-Industrialization Objects in the Russian Economy

Olga Brizhak[1(✉)] and Oleg Tolstobokov[2]

[1] Department of Economic Theory, Financial University Under the Government of the Russian Federation, Moscow, Russia
brizhak71@mail.ru

[2] Moscow Technical University of Communications and Informatics, Moscow, Russia

Abstract. In the study, we used a modern version of the factorial analysis of neo-industrialization objects in the Russian economy. Industrial enterprises of old and new types, including those transformed due to neo-industrialization, were chosen as neo-industrialization objects. Effective functioning of industrial enterprises is known to ensure high rates of sustainable national economic growth, independence from foreign markets, economic strength and stability. It is apparent that successful integration of industrial enterprises into the new economic reality requires the development of industry and its industrial core. We performed a factorial analysis of industrial enterprise operation, revealed the role and impact of the human factor in Russian industry, developing on a new technological basis that corresponds to the new technological mode, determined the significance of the intellectual potential of the human factor at the enterprise, and revealed the role and significance of the organizational factor. It is obvious that restoration of the capacity of the Russian economy, modernization of its industrial complex, dissemination of innovation, and implementation of the import substitution program, especially in the context of emerging geo-economic and geopolitical challenges, are of particular relevance in current conditions.

Keywords: Factors of Production · Enterprises · Neo-Industrialization · Technological Breakthrough · Human Capital · Russian Economy

1 Introduction

The current stage of the new era is marked by fundamental changes and contradictions in social and economic life, accelerated pace of technological advancement, and transition to a new technological structure. In the study, we emphasize the relevance of neo-industrialization of the domestic economy, which is being successfully implemented in the real sector of the economy and the high-tech sector, despite the pressing development challenges, reversal of trends, and confrontation of heterogeneous forces. Fierce global competition for leadership is growing, environmental problems are worsening, and the struggle for the planet's resources and markets is escalating. The role of enterprises in the real sector of the domestic economy under these conditions is growing. Modern enterprises must be flexible, mobile, and promptly adaptable to dynamic

R. Polyakov (Ed.): EcoSyst 2023, LNNS 1092, pp. 14–21, 2024.
https://doi.org/10.1007/978-3-031-67354-2_2

external environment. In conditions of fierce competition, some enterprises suffer from inertia and excess functions many of which are in low demand [1]. Flexibility and adaptability are characteristic of clusters that are able to solve complex problems of modern development. The centers of industrial clusters are large technology corporations, innovative universities, etc. Such centers of gravity are surrounded by a cluster periphery represented by numerous small enterprises. Small enterprises provide services in data analysis, factual material processing, project development, consulting, financial support, accounting and auditing, building communications, and advanced training and retraining of personnel. A large enterprise owns a set of unique resources, produces a certain type of goods, keeps in possession capital, technology, possesses organizational capabilities and management competencies, and has its own needs and interests. The state of large enterprises, like individual entities, can be different: from stability to crisis, from booms to downturns; it can be a state of depression or revival; they can be involved in external processes or remain aloof from current events, or be in a state of expectations of future changes. Market valuation of large enterprises and corporations is sensitive to such states, and therefore it either increases or drops [2]. Factorial analysis of enterprises sheds light on how well the organizational and management mechanisms of the company are structured, how unique the capital combinations of production factors are, and on the role and value of intellectual potential of the company.

2 Methodology

The factorial analysis of neo-industrialization objects of the Russian economy cannot be performed within narrow boundaries; it should be systemic, it requires interdisciplinary recourses and worldview analysis. The Russian economy is primarily aimed at improvements in the qualitative composition of the enterprises to produce high value-added goods in the technological sector of industry and in production of modern personal goods. Their share in the Russian economy is still insignificant. The search for means to promptly elaborate the solution showed the need for the idea of neo-industrialization and accretion of its followers [3]. This idea implies creation of technology-based industrial sectors to produce high value-added goods. The Russian economy makes neo-industrialization its imperative to efficiently use the accumulated rental income for reconstruction of the national manufacturing industry based on advanced technologies.

In recent years, neo-industrialization caused significant changes in the industrial enterprises of the Russian economy and affected its budget, reserves and policy. Enterprises operating in newly created manufacturing industries can produce high value-added goods and become the key participants in the domestic economy. They determine its global competitive position and investment attractiveness, and attract the intellectual potential of Russian society. Enterprises operating in extractive industries and producing low value-added goods gain access to advanced technologies and equipment, which inevitably reduces the level of technological dependence of the country and improves its global position [4].

At the same time, the dependence of the government budget on the flow of rental income decreases, and the contribution of innovative production to the state budget grows. National reserves are converted into funds allocated for modernization and innovative development of the economy.

2.1 Assessment of the Transformation and Development of Russian Enterprises in the Russian Economy in the Context of Neo-Industrialization

The market economy of the era of true capitalism involved numerous market participants, and the share of each participant was negligible to have a significant impact on the market. Small market participants relied on their own production factors and made their decisions on issues of production, management and marketing of products. However, the global situation has changed, and the current economy is dominated by large enterprises, corporations, and unions they form with their own market share; they set the 'agenda' in the modern social economy.

A large enterprise owns a set of unique resources, produces a certain type of goods, keeps in possession capital, technology, possesses organizational capabilities and management competencies, and has its own needs and interests.

Russian enterprises have passed a long way of formation and development, adapting to different areas of activity and changing their form. Each enterprise has its own life cycle and goes through a number of phases from foundation to liquidation. Each enterprise goes through its own life cycle, gains production and market experience, competes, and attains a certain position in the market. For example, enterprises operating within a command-administrative economic system corresponded to the economic system of socialism with a social system and a bureaucratic vertical and labor collective with plan tasks assigned from this vertical. Fulfillment of the plan ensured the enterprise with income and government support. Government officials assessed activities of the enterprise, since the government and not the market was the key production coordinator [5].

The stage of restructuring and privatization of state property at the end of the 20th century was followed by the period of formation of integrated entities: corporations, holdings, financial and industrial groups, etc. It took decades to attain successful formation, functioning and stabilization of new organizational and legal forms. Yet hidden processes currently occurring in some enterprises make them nontransparent, question their investment attractiveness, and affect their competitiveness.

In current conditions, enterprises must be ready to face up global competition, which is incompatible with the groups within the enterprise that 'appropriate' its functions and redistribute a significant part of development resources for their own benefit. Readiness for global competition in current conditions can be ensured through effective public control over the activities of the largest Russian corporations, such as Gazprom, Rosneft, Rosatom, and Russian Technologies, and through significant changes initiated in the internal environment of such corporations [6], (Fig. 1).

The system of goals that set the course of national policy increasingly frequently puts forward qualitatively new guidelines: the leading positions of Russian corporations in the World Competitiveness Ranking; expanded reproduction of intellectual capital; increased life cycle of enterprises. The Russian economy exhibits a high degree of diversification – it is not only the economy of the fuel and energy complex, but also the economy that effectively employs other production factors, such as information, technology, and intellectual factors.

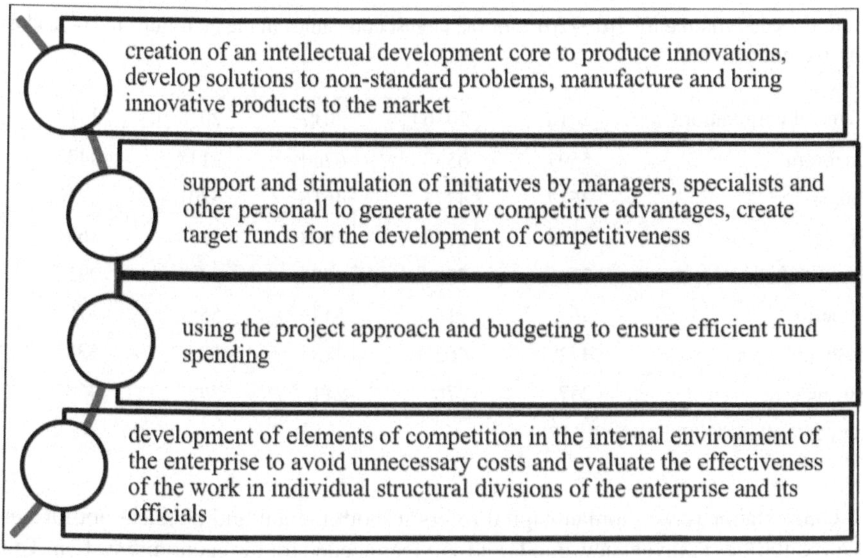

Fig. 1. Formation of basic changes in the internal environment of Russian corporations. Source: Compiled by the authors based on research data

In the ranking of the largest industrial corporations, the total sales volume is formed by more than half due to 29 companies in the petroleum and oil and gas industry, which is 14.3% of the number of industrial companies (Table 1).

The most significant in terms of sales volume are transport companies which accounts for 6.9% of the total sales volume of 600 companies. The other 144 companies from such industries as telecommunications and communications, energy sales, engineering, construction, agriculture, information technology and others that account for almost a quarter of the total number of companies included in the rating show insignificant sales volumes. The sales volume of the largest industrial companies in the mechanical engineering, ferrous and non-ferrous metallurgy, electrical engineering, and chemical and petrochemical industries attains 35.1%. Over the past two years, the number of high-tech enterprises has increased 1.5-fold owing to effective implementation of the neo-industrialization policy in the Russian economy [7].

2.2 Human Capital as a Key Production Factor in Neo-Industrialization Processes

Over the past decades, human capital has become a powerful alternative to the labor power and a key factor in the high-tech economy. Human capital implies that not a thing but a living person who is part of the organizational factor that interconnects other factors (technological, financial, etc.) of production at one enterprise is the factor of value. Human capital is closely related to its owner and creates value as part of the owner. In contrast to labor force, human capital features qualities like work experience, social skills, and qualifications. When lacking means of subsistence, human capital with its opinions, rights and economic interests starts to reproduce the ability to work [8].

Table 1. Sales volume for 2015–2019 for the largest companies in the Petrolium and oil and gas industry, billion rubles

Name of corporations	2015	2016	2017	2018	2019
Gazprom	5597	6537	6764	7145	8477
Lukoil	3617	4275	4616	5801	6740
Rosneft	2373	2858	3098	4176	4681
Surgutneftegaz	722	789	950	938	991
Tatneft	568	716	544	555	676
Bashneft	472	462	488	503	538
Novatek	257	276	211	398	458

Source: Ranking-600. Expert-RA[7]

Unlike labor force, human capital refers to both mental and physical abilities and creative abilities to invent new goods and services needed for the economic system. Thus, human capital exhibits a higher stage of development of the personal factor of production. Labor force is the initial stage of labor development, which is the employer's object of sale and purchase needed to start production and gain a profit based on the developed abilities that are in demand in the market, free to use personal powers, and not involved in other areas to ensure a life of material comfort.

The imperative of neo-industrialization creates additional needs for human capital as a unique factor in creating innovation for high-tech production. Russian society has many built-in restrictions on the formation of unique individual abilities and their capitalization in human capital. Hence, neo-industrialization policy primarily focuses on formation of the intellectual component of human capital [9, 10]. Figure 2 presents the key components of the triune task of human capital formation.

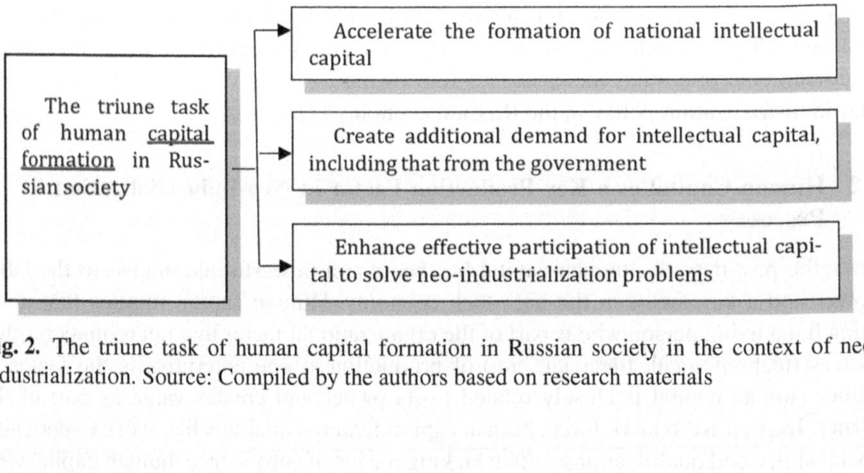

Fig. 2. The triune task of human capital formation in Russian society in the context of neo-industrialization. Source: Compiled by the authors based on research materials

In the context of neo-industrialization, intellectual and creative work is expected to predominate in sectors that provide breakthrough results in the most critical directions of the social system.

It should be noted that human capital and its productivity largely depend on implementation conditions. A higher human capital and better conditions for tangible embodiment ensure faster and more efficient transition from one phase of human capital reproduction to another. In case of the lack of prospects for development and conditions for human capital implementation, a person becomes passive; this is manifested in the quality of its passage through the phases of the reproduction process [11, 12].

The formation of national human capital starts with the stages of early identification of different functional forms of this capital and allocation of targeted funds from the government and public organizations in the form of grants or other payments. Figure 3 illustrates the stages of human capital reproduction in the context of neo-industrialization.

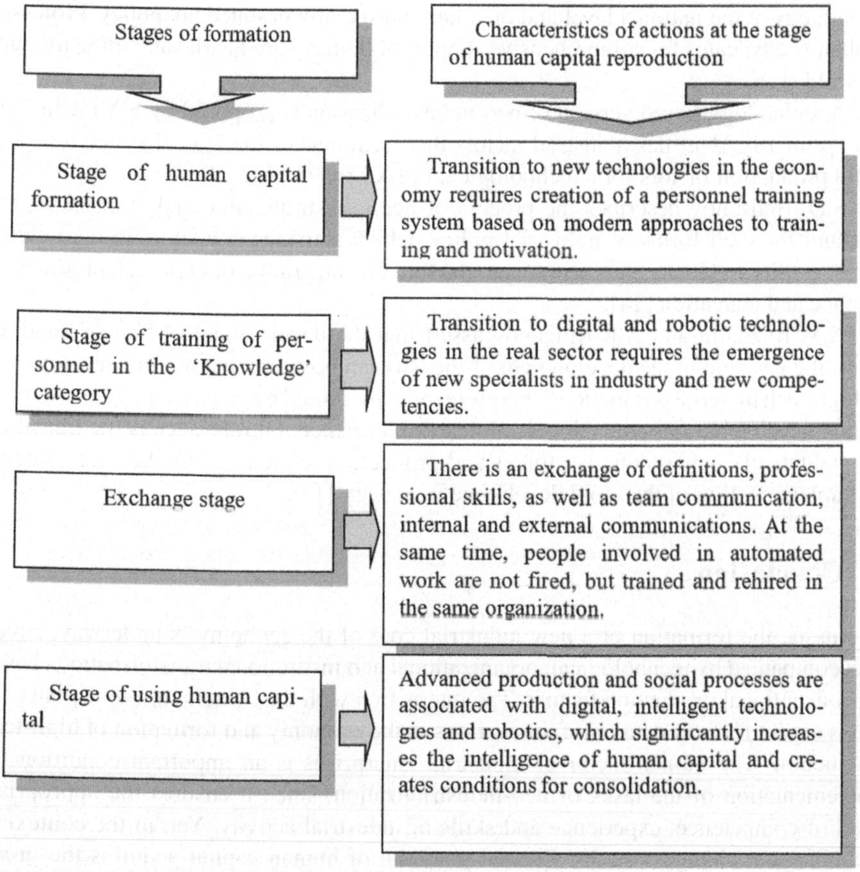

Fig. 3. Human capital reproduction in the context of neo-industrialization. Source: Compiled by the authors based on research materials

As can be seen from Fig. 3, the stages of human capital development are designed to change the current economic trends in Russia. The key development imperative in the context of neo-industrialization is a technological breakthrough to ensure the global level of competitiveness of the country and increase its influence on the global stage. The competitive advantage of the country in the modern world depends on the level of knowledge, technology, and competencies of professionals, and flexibility of organizational mechanisms and institutions established by the creative forces of free intellectuals.

3 Result and Discussion

The topic of the study is debatable, and a number of authors have focused on this topic over several years. We agree or debate with these authors, or share their concepts. One of the leading concepts in the field of neo-industrialization and the use of factor combinations of the enterprise is the concept by Professor S.S. Gubanov, who indicates the urgent need for a systematic approach to this issue, primarily focusing on personnel and technology at the national level and on a fair and socially oriented tax policy. Professor Gubanov advocates for comprehensive support of domestic industrial and infrastructure capital [3].

A technology-based version of neo-industrialization is proposed by S.Yu. Glazyev, who primarily identifies industrial factors that form new technological structures – the main production factors for a technological breakthrough [13].

S.D. Bodrunov describes the process of neo-industrialization and emphasizes the demand for such forms of industrial policy, which activates regulators in the field of taxes, credits and loans, thus bringing to the fore the integration of high-tech production, science and education [14].

A.V. Buzgalin and A.I. Kolganov assert that the development of human qualities plays the key role in the development of the Russian economy but not the development of high-tech material production. Therefore, priority should be given to such areas of economic development as education, healthcare and science. Of relevance is the transition to qualitatively new productive forces and production relations as the key components of implementation of the rapid development strategy [15].

4 Conclusion

In Russia, the formation of a new industrial core of the economy is underway, which is accompanied by technological, organizational and institutional transformations based on conventional production factors. The interaction with the owners of intellectual capital is urgent for introduction of innovations in the economy and formation of high-tech production. The human factor of industrial enterprises is an important condition for implementation of the tasks of neo-industrialization, since it ensures the appropriate level of competence, experience and skills of industrial activity. Yet, in the context of technological changes, the intellectual potential of human capital acquires the greatest significance, since it is capable of generating new ideas, developing new capital combinations, participating in industrial communications, and producing innovations that are necessary for the reconstruction of domestic industry on a new technological basis. An adequate response to the challenges of modern development is the implementation of technological, structural and socio-economic transformations of both the neo-industrialization objects and the economic system as a whole.

References

1. Brizhak, O.V., Tolstobokov, O.N.: Forming the new industrial core of Russian industry: problems and perspectives. Res. Econ. Anthropol. **42**, 155–160 (2022)
2. Polyakov, R., Brizhak, O.: Industrial clusters and the process of their self-organization. Lect. Notes Netw. Syst. **705**, 60–72 (2023)
3. S. S. Gubanov. Powerful breakthrough. Neoindustrialization of Russia and vertical integration. Moscow: Knihniy Mir (2012)
4. Brizhak, O., Adamenko, A.: Neoindustrial paradigm of russian corporations integration into the economy system under the conditions of transition to the sixth technological order. Lect. Notes Netw. Syst. **111**, 619–626 (2020)
5. Brizhak, O.V., Chikanova, E.S.: From previous economic order to new economic reality. Lect. Notes Netw. Syst. **372**, 347–353 (2022)
6. Kleiner, G.: Socio-economic ecosystems in the light of system paradigm. In: System analysis in economics – 2018. Proceedings of the V International research and practical conference-biennale, pp. 4–12 (2018)
7. Industry structure of the RAEX-600 ranking (by sales volume). www.raexpert.ru/project/top companies/2021/ranking
8. Skokov, R.Y., Brizhak, O.V.: Consumption of addictive goods in Russian regions and its impact on the quality of human capital. R-Economy **6**(1), 50–60 (2020)
9. Brizhak, O., Polyakov, R.: Creative impetus for the development of innovation clusters of the Russian economy. Lect. Notes Netw. Syst. **474**, 99–104 (2022)
10. Shinkevich, A.I., Shinkevich, M.V., Malysheva, T.V.: Management of the process of integration of economic entities into innovation-oriented competitive clusters. Manag. Econ. Syst. Electron. Sci. J. **2**(96), 2 (2017)
11. Brizhak, O.V., Romanets, I.I.: Creative potential in the development of national ecosystems. IOP Conference Series: Earth and Environmental Science. International conference Germany and Russia: Ecosystems without borders **689**, 102–107 (2021)
12. Barkhatov, V., Benz, D., Pletnev, D.: Trust in the Russian economy and business environment. E3S Web Conf. Innovative Technol. Sci. Educ. **210**, 13013 (2020). ITSE 2020
13. Glazyev, S., Ajvazov, A.E., Belikov, V.A.: The future of the world economy is an integrated world economic structure. Econ. Regions **14**(1), 1–12 (2018)
14. Bodrunov, S.D.: Reindustrialization in the conditions of a new technological revolution: the road to the future **10**(5), 2–8 (2019)
15. Kolganov, A.I., Buzgalin, A.V.: Reindustrialization as nostalgia? Theor. Discourse. SOCIS **1**(357), 80–94 (2014)

Regional Innovation Ecosystem Management Principles in Order to Implement the Structural Transformation of Innovative Processes in the Region

Elena Kirillova(✉), Andrey Shendrikov, and Sergey Merkushev

Smolensk Branch of the National Research University «Moscow Power Engineering Institute», Smolensk, Russia
kirillova.el.al@yandex.ru

Abstract. One of the ways to achieve sustainable growth in the strategic perspective is the priority of transforming industrial and socio-economic processes based on innovation, the implementation of which is impossible without close interaction and co-development of individual participants in their creation, as well as infrastructure elements, which determines the need to apply ecosystem principles to manage them. Attempts to develop individual participants and supporting elements of such complex systems autonomously, point-by-point, without taking into account, coordinating and balancing the interests, capabilities and limitations of each of them slow down the achievement of the strategic orientations of individual participants, the priorities of the association and negatively affect their sustainability. Based on the analysis of the structural transformation of innovative processes in the region, the article describes their identified key features, trends and directions of change. Based on this specificity, the requirements for the development of approaches and management methods for achieving the objectives of these formations and realizing their advantages in a strategic perspective are formed. The developed principles of regional innovation ecosystem management are described, which should be taken into account when forming a methodology for the creation and development of these systems in the context of structural modernization of modern complex socio-economic and industrial systems based on an ecosystem approach. Taking into account the formulated methodological principles in the development and practical application of management tools will ensure the achievement of sustainable development both for themselves and for receiving benefits from other participants and the system as a whole in a strategic perspective.

Keywords: Principles · Innovation Ecosystem · Regional Innovation Management

1 Introduction

The process of creating an innovative product and its commercialization is quite complicated from the point of view of management due to the complexity of its description, weak formalization, difficulties in identifying objects of influence, calculating its strength, as

R. Polyakov (Ed.): EcoSyst 2023, LNNS 1092, pp. 22–32, 2024.
https://doi.org/10.1007/978-3-031-67354-2_3

well as forecasting and evaluating the results, and the time of its receipt. At the same time, limited resources, the need to ensure the direction of managerial influences in order to achieve the goals and objectives set by the production and economic entity, as well as the need to "adjust" other participants and elements of the regional ecosystem to its speed, capabilities, prospects and limitations of functioning determine the relevance of the formation of basic general provisions or basic rules that can be formulate and establish a framework for such activities. Under the management of this type of system, we will understand a set of targeted measures of flexible and adaptive impact on the continuously evolving integration of production and economic entities, the totality of their connections and infrastructure elements, as well as specially organized provision of opportunities for the implementation of such activities aimed at achieving sustainable competitive advantages for the system as a whole and benefits for its participants in a strategic perspective.

Despite the presence of a significant number of studies affecting the theoretical, methodological and practical aspects of managing various forms of integration, ensuring the implementation and development of innovative processes at the regional level, there are practically no theoretical proposals for characterizing, identifying and describing the dynamic abilities of a cooperative group of interacting entities localized in a certain territory and aimed at maximizing the use of their own and regions' internal resources opportunities for catalyzing the initiation of innovative ideas, to support and consolidate their implementation. While in modern conditions of high turbulence and weak predictability of external conditions, the use of external points of support for choosing the vector of development of an industrial and economic entity, as determined within the framework of a positional approach to strategic management, is rather not advisable. It seems promising to focus on the provisions of the concept of dynamic abilities, which is an evolutionary development of the resource approach and is based on the identification, enhancement, and catalyzation of resource combinations of participants' potentials through their flexible integration into value chains and obtaining synergetic, cumulative, and multiplicative effects.

2 Structural Transformation of Modern Production and Economic Systems

Structural transformation, which is always inherent in the process of development by its nature, has been significantly accelerated recently and covers an increasing number of subjects, phenomena and processes. There is also a decrease in the predictability of these changes and an increase in the "price of error" [1, 2], which determines the urgent need to choose approaches and, based on them, form methodological and practical management tools for modern industrial and economic entities that not only take into account these changes "at the moment", but are also able to adaptively and flexibly rebuild, while maintaining a strategic focus on the mission, focusing on the basic general rules and targets.

Macro and micro changes, of course, are primarily reflected in the most flexible and adaptive socio-economic and industrial structures. It is possible to note the transformations of individual subjects, their priorities and, in accordance with them, behavioral models. Competitive market mechanisms in response to changes in consumer demands and macro trends have led to quite large-scale transformations at enterprises in the real sector of the economy in recent years. Modern industrial and economic entities and the totality of their connections are gradually transforming into ecosystem forms of interaction that form collective network competencies and give them the ability to continuously update. The same large and rapid transformations are required in the field of training specialists in order to support the implementation of all changes, since the creation, effective functioning and development of such forms is impossible without the co-development and balance of interests of all their participants and elements.

To increase the intensity and effectiveness of these structural transformations, it is necessary to restructure the ways of organizing such activities in the direction of more flexible and adaptive forms, where informal networks play a special role. Network collaborative forms of organization of interaction between participants in production and socio-economic systems are beginning to play roles that previously played the industry. The sharp technological leap has had a strong impact on the entire structure. There is a significant increase in the role of modern digital technologies, high-tech industries and industries, innovative solutions in which radically change the structure of industries, as well as business models, factors and conditions of competitiveness of individual subjects, regions and countries. In the process of creating and commercializing innovations in the market, it is now increasingly not individual enterprises that compete, but value chains involving other economic entities of one or even several regions, as well as elements of regional infrastructure. The formation and management of such a system of interrelated processes leads to the need for process approaches to innovation management, for example, logistics. Its application is revealed in the following principles: minimization of logistical costs; information management; synergetic effect; parametric management; systematic management [3].

During the analysis of the practices of formation and development of innovation systems in Canada, India, Brazil in [4], the author highlights such general principles that are proposed to be used for developing countries: innovation is the basis of competitiveness; greening the economy; cooperation; globalization; priority development of regions; mutually beneficial cooperation; active stimulation of innovation; support for all participants in the system.

The management of the innovation process within the framework of the national economy takes place in conditions of an increasing role of territorial factors and a shift in the emphasis of innovative development to the regional level. Therefore, special attention should be paid to the construction of a regional innovation system [5], including to take into account the peculiarities of territorial economic interests, the specifics and level of development of the region, and the ratios of technological structures of the economy. The development of regional innovation systems should be based on integrated structures functioning in accordance with the following principles: continuous improvement of the quality of human capital; increased investment in education; increased innovativeness of entrepreneurial activity; expansion of sales markets for high-tech products through business contacts in interregional and international markets; effective risk management; integration and synergy [6]. At the same time, the orientation and use of only the principles of management of regional innovation systems in the formation and development of regional innovation ecosystems is impractical, since these formations are more complex and have certain specific features.

3 Methodological Principles in Accordance with the Structural Dynamics of the Transformation of Innovative Processes in the Region

Due to the new challenges faced by modern production and economic systems, the transformation of their elements and the systems themselves, the principles of their management should also take into account these changes.

The dynamics and direction of change of the innovation process itself determines the principles of flexibility, openness, adaptability. The implementation of open innovative practices requires the formation and support of the development of appropriate ecosystem forms of organization of participants and infrastructure elements. The article [7] notes that innovations change the structure of production and economic systems significantly and the key feature of the effectiveness of modern forms is the intersectoral ecosystem character. One of the ways to achieve sustainable growth in the strategic perspective is the priority of transforming industrial and socio-economic processes based on innovation, the implementation of which is impossible without close interaction and co-development of individual participants in their creation, as well as infrastructure elements, which determines the need to apply ecosystem principles to manage them.

At the same time, innovation is a rather complex phenomenon that is difficult to identify, describe and measure, including due to its interdisciplinarity. The complexity of production technologies, works and services increases the requirements for the qualification of specialists, as well as narrows the scope of application of these special knowledge and skills for each of them. An employee with a narrow specialization "alone" cannot create a complete high-tech product, job or service, since most modern innovative tasks are complex and large-scale, which necessitates cooperation.

The increasing role of intellectual property in the sustainable development of modern industrial and economic entities is of significant system-forming importance for such systems within the framework of the implementation of the interaction of their elements.

The innovation process is now perceived as continuous. It is often presented as a sequence of certain stages interconnected with each other and leading to the generation of new ideas, proposals, innovative knowledge, commercialized in the form of a product, work or service. Innovations are considered not as a linear sequence of operations, but as a dynamic multidimensional process, which determines the allocation of such specific features of the environment as accessibility, openness, flexibility [8]. These needs have become possible to realize with the development of information and communication technologies [9].

According to the methodological recommendations of the EU [10], for systems of this kind, the inseparability and importance of relying on identification and creating conditions for the possibility of revealing their uniqueness by each of the participants and elements, creating due to their diversity that unique "cocktail" of competitive advantages realized through cooperative partnership, is emphasized. The principle of concentrating managerial influences and resource capabilities to create a critical mass of diverse and complementary elements is one of the key ones.

With the significantly increasing role of knowledge in explicit and implicit form, the share of intellectual activity results in the value of enterprises [11], as well as taking into account the tasks set for domestic industrial enterprises for accelerated modernization for import substitution without loss of quality, as well as preventing the disruption of production chains of goods, works and services [12], it is the provision of human resources that comes out to the fore.

The role of innovations, approaches and mechanisms of their management in the framework of the development of the localization territory have been considered in sufficient detail and now more and more authors determine that achieving sustainable development goals is possible when considering them within the framework of ecosystem formations [13]. This approach emphasizes the common focus of interaction between regional actors and elements, their interconnectedness and interdependence. They complement and enrich each other, ensuring a proportionate and balanced innovative development of all participants in the system, thereby contributing to the development of national innovation systems.

The emphasis on the priorities of sustainable development and the provisions of the closed-loop economy is noted in [14] when formulating the principles of creating ecosystem cyclical innovations, understanding cyclicity as a key property of the system, and not as a property of a separate product or service. Based on it, the authors identify three groups of principles: collaboration, experimentation and platformization. Modern researchers of the ecosystem approach emphasize such a fundamental feature of ecosystem creation as ensuring the possibility of unhindered consumer access to a variety of services [15], when the consumer gets the opportunity to minimize the cost of searching and comparing and can receive the entire personalized set of resources, works and services in one place.

To manage the implementation of such interaction, the authors propose to rely on the concept of dynamic abilities, which focuses on intellectual assets as the main sources of long-term sustainable benefits, the main drivers of growth and profitability in most sectors of the economy [16]. When defining knowledge formed within the framework of ecosystem interaction between industrial enterprises and scientific and educational organizations with the support of public authorities as a valuable resource of a modern subject of industrial and economic relations, the logic of its strategic actions becomes a continuation of the resource approach in general and the concept of dynamic abilities in particular [16].

4 Results

Principles that contribute to the formation of collective network dynamic abilities and give them the ability to continuously update within the framework of the structural transformation of innovative processes in the region.

In the context of rapid technological changes, according to the concept of dynamic abilities, subjects in a strategic perspective need to focus on competitive advantages as the basis for identifying distinctive processes (methods of coordination and combination) determined by (specific) asset positions. At the same time, the important principles are to support the improvement of internal technological, organizational and managerial processes [17]. The development of organizational and human resources potential, accumulation of competencies, creativity and idea management, culture and climate, as well as technology management as sources of wealth creation, rather than ownership of physical assets, are noted as priority principles in [18].

The focus on the development of new products and processes in combination with business model innovations for the use of production and economic entities of their dynamic capabilities is noted as a key principle in management [19]. At the same time, the ecosystem approach of obtaining the entire range of goods, works and services in one place at one time based on digital technologies [20] significantly strengthens it.

Modern industrial and economic entities are guided by the principle that creating values for all their stakeholders is an integral component of success. The strategic complementarity of innovative projects is emphasized in [21]. This means that the simultaneous implementation of all these projects provides a great advantage, which leads to even more active transformative activities. At the same time, the urgent need for radical transformations of approaches and principles of managing innovation processes in the regions is emphasized. Such transformations should be aimed at significant transformations of existing structures, such as accelerating innovation to solve large-scale social problems or modernizing traditional sectors of production and economic systems [22].

Thus, methodological principles in accordance with the structural dynamics of the transformation of innovative processes in the region contribute to the formation of collective network dynamic abilities and give them the ability to continuously update (see Fig. 1).

Within the framework of managing the interaction of individual industrial and economic entities in the implementation of innovative processes at the regional level, more and more attention is paid to predicting the transformation not of themselves and the environment, but to identifying and evaluating the dynamics of changes in the system as a whole with the co-development and mutual influence of constituent elements. Participants in the system of industrial and economic relations play complementary roles, transforming and evolving together. The complexity and cross-cutting nature of the processes of creation and commercialization of innovative products, works, services and structural interconnectedness, interdependence of individual participants, their divisions, and elements among themselves only strengthen these trends. With this in mind, the goal of developing an innovative ecosystem is not to make a one-time profit, but to ensure structural transformations to create conditions for a continuous steady stream of positive changes in the strategic perspective through the implementation of the principle of coevolution. The principle of coevolution can be described as the coordinated development of all participants in regional innovation processes, supported and in some cases coordinated by the innovation ecosystem. At the macro level, these actions, in accordance with these rules, determine the stability of the ecosystem, its participants, the region of localization and social consistency.

The openness of modern systems and the continuity of current changes determine the cyclical nature of the innovation process, where a sequence of transformative actions leads to the achievement of planned values and, at the same time, to an increase in the scale of transformations. This repetitive cycle of changes ensures close interaction of

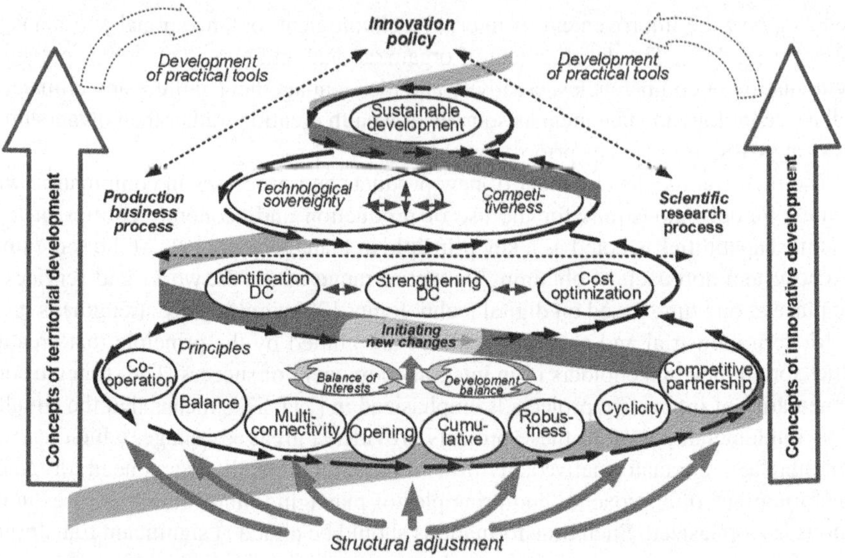

Fig. 1. Methodological principles in accordance with the structural dynamics of the transformation of innovative processes in the region.

all subjects connected within the innovation ecosystem, consistently redistributing the main attention in favor of data collection and prioritization in actions for planning and implementing changes in them, in evaluating the results of such transformations [16]. The principle of cyclicity implies the implementation by the innovation ecosystem of a process approach to innovation, in which the results of innovative projects should be further used as the basis for the next cycle of development of the regional innovation system.

Structural shifts also determine changes in the vectors of influence of the main market forces. Cooperation in the current conditions is not a sign of a lack of competencies or resources from one of the participants. In this case, it is understood as a strategic decision on partnership for optimal use of the capabilities of subjects, territories and markets to improve the position of each of them in the competitive struggle. In this case, the goal of the innovation ecosystem is to form a certain community of innovatively active enterprises and innovative infrastructure organizations that create new value through a balance of cooperation and competition, which can be implemented based on the principle of competitive partnership. At the same time, this principle emphasizes the selection and survival in the long term of the necessary and sufficient diversity of participants and elements of the ecosystem.

The concentration of attention, limited resources, and core competencies of individual entities only on key business processes increases the need for coordination and coordination of the interests of individual participants, the interconnection of their goals and capabilities into a single value chain. The principle of balance presupposes the orientation of the innovation ecosystem towards resource and organizational provision of a balance of interests and capabilities of participants in the chain of creation of innovative products to achieve the goals of socio-economic development of regions and the country as a whole. On the part of government agencies in their changing role, it allows ensuring the budgetary effectiveness of the measures taken. The implementation of this principle forms the basis for the adaptability of individual participants, their interrelationships and formation as a whole, taking into account the constantly changing needs of all elements and the environment, which are constantly preparing, accepting changes and constantly "adjusting" to them. At the same time, it should be noted that it cannot be implemented without parallel work on the formation and development of the level of trust between the participants, their flexibility and openness, as well as the distribution of areas of responsibility, a system of "checks and balances", establishing the necessary level of communication and interchange.

The rather narrow specialization of modern industrial and economic entities within the framework of one of the functions or stages of the innovation process, together with the possibilities of implementing information and communication interaction represented by digital technologies, their complementary roles in the structure of the region cause such a sharp and rapid leap in the development of networks, as well as a certain dependence on them. The principle of multi-connectivity (or network nature) determines the expediency of the implementation by the innovation ecosystem of the function of transferring the results of the stages of innovative processes, as well as resources of various types between participants in these processes within the framework of the network structures being formed. This principle is implemented in the total involvement and interconnection of all spheres of activity of both individual participating enterprises and themselves as a whole in this process. These structures should include coordinating centers (nodes) and ensure the selection of optimal (in terms of timing, costs and results) routes for patent flows. At the same time, the ecosystem takes into account various possibilities of changing the state of individual participants, elements and the external environment, implementing requirements to ensure its dynamism and flexibility, i.e. a sufficiently rapid response to any change in the state of the managed object or the external environment and the ability to rebuild under the influence of these changes.

With the help of an innovative ecosystem, the formation of open networks of mutual exchange and the development of general and special knowledge based on mutually beneficial relationships between individual participants allows you to quickly respond to changes in the implementation of innovative processes. The Internet and other digital opportunities only reinforce these trends. The principle of openness characterizes the controlled interaction of participants in the innovation process both among themselves and with the external environment in order to maximize the potential of open innovation based on the balance of "openness/protection of the patent".

In the structure of the modern production and economic system, the share and strength of the influence of knowledge networks aimed at accumulating and catalyzing its potential over time is gradually increasing. Also, through continuous interactions of ecosystem participants for its internal and external innovative transformations, this potential for change is enhanced and becomes a source of new ideas and proposals, creating an information flow of patent, which forms the innovative potential of the region based on elements of the digital economy. This defines the principle of cumulativeness, according to which the innovation ecosystem should provide information, organizational and technological support for the catalysts of change that produce innovations. This principle is implemented by solving the problem of global optimization to achieve synergetic effects with the general use of its individual elements. Such optimization should be aimed primarily at the costs caused by resistance to change and preventive measures to minimize them.

The regional innovation ecosystem is characterized by a large number of parameters, characteristics of subjects, and connections between them, many of which are defined vaguely or in qualitative formats. This determines that when managing it, it should be based on the principle of robustness, which involves the formation of some approximate models of the control object with the allocation of key parameters on which the impact is carried out to achieve the goals of innovative development. This is a complex principle that assumes the stability of the functioning of the innovation ecosystem with sharp (in general, poorly predictable) deviations of the indicators of participants in the innovation process and external disturbances from the predicted indicators.

5 Conclusion

Thus, the principles of managing the regional innovation ecosystem of the region, formulated on the basis of the highlighted features of the structural transformation of modern innovative processes in the region, can become the fundamental rules governing the creation and development of these formations in modern conditions. Their application as a flexible conceptual framework will allow accumulating the resources and abilities of individual subjects, ecosystem elements and their localization territories, identifying and contributing to development through concentration and complementarity of their core competencies, thereby directing end-to-end processes for the creation of innovative products and their commercialization, optimizing costs, as well as ensuring the achievement of set targets for individual the subjects of interaction and the system as a whole.

The need for their availability and the formulation of principles in this form are due to the identified and described key features of the structural transformation of innovative processes in the region, trends and directions of their changes. The process of creating an innovative product and its commercialization is quite complex from the point of view of management due to the complexity of its description, formalization, allocation of objects of influence, calculation of its strength, as well as forecasting and evaluating the results, the time of its receipt. At the same time, limited resources, the need to ensure the direction of management actions to achieve the goals set by the production and economic entity, as well as the need to "adjust" other participants and elements of the regional ecosystem to its speed, capabilities, prospects and limitations of functioning determine the relevance of the formation of basic general provisions or basic rules that can be formulated and established for such an activity. At the same time, in modern conditions of high turbulence and weak predictability of external conditions, it seems promising to focus on the provisions of the concept of dynamic abilities, which is an evolutionary development of the resource approach and is based on the identification, strengthening, catalyzation of resource potentials through their flexible integration into the adaptive formation of value chains and obtaining synergetic, cumulative and multiplicative effects.

Methodological principles, in accordance with the structural dynamics of the transformation of innovative processes in the region, contribute to the formation of collective network dynamic abilities and give them the ability to continuously update For each level of integration, the region of localization of the innovation ecosystem, the described principles will be concretized in a set of practical tools and measures reflecting the specific features of the object of managerial influence and revealing their abilities, maximizing added value.

Acknowledgments. The research was carried out at the expense of a grant from the Russian Science Foundation (project No. 23–78-01197).

References

1. Kirillova, E.A., Malevich, E.P.: Analysis of the structural features of the production and economic system of the region in the aspect of the practice of introducing information elements to support the implementation of innovative processes. Soc. Politics Econ. Law **8**(121), 129–135 (2023)
2. Zaenchkovsky, A.E., Lazarev, A.I., Ledneva, O.V., Prokimnov, N.N.: Analysis and testing of algorithms for neural network TCP/IP packet routing in private virtual tunnels. Appl. Inf. **6**(108), 98–108 (2023)
3. Zaenchkovsky, A.E.: Fundamentals of logistics of integration processes in innovation-oriented structures. Sci. Rev. **6**, 614–618 (2012)
4. Kiseleva, O.N., Sysoeva, O.V., Vasina, A.V., Sysoev, V.V.: Updating the open innovation concept based on ecosystem approach: regional aspects. J. Open Innovation: Technol. Market Complex. **8**(2), 1–30 (2022)
5. Zaenchkovsky, A.E.: Methodology of logistic and organizational management of the innovative environment of regional scientific and industrial complexes. Russ. Entrepreneurship **14**(236), 129–137 (2013)
6. Troshin, A.S., Stolyarova, Z.V., Lyapunov, S.A.: Practical implementation of an effective mechanism of integrated structures through the construction of a regional innovation system. Innovations Investments **2**, 235–241 (2023)
7. Gamidullaeva, L.A., Tolstykh, T.O.: Implementation of cross-industry projects on the principles of ecosystem as a new vector of innovative development. Innovations **8**, 65–74 (2020)
8. Kakatunova, T.V., Khvostov, D.A.: Strategic management of innovative development of the region. Secur. Probl. Russ. Soc. **3–4**, 140–145 (2014)
9. Wang, F., Stoianova, O.V.: Modeling the effect of e-service quality and knowledge culture on employee performance. J. Appl. Inf. **5**(107), 5–19 (2023)
10. Regional CoR guide: Innovation Ecosystems, p. 272. European Committee press, European Union (2016)
11. World Intellectual Property Report: The direction of innovation, p. 98. WIPO, Geneva (2022)
12. Oruch, T.A.: Research of indicators and results of import substitution in the Russian industry. Innovations Investments **1**, 289–293 (2023)
13. Gamidullaeva, L., Shmeleva, N., Tolstykh, T., Shmatko, A.: An assessment approach to circular business models within an industrial ecosystem for sustainable territorial development. Sustainability **2022**(14), 2 (2022)
14. Konietzko, J., Bocken, N., Hultink, E.J.: Circular ecosystem innovation: an initial set of principles. J. Clean. Prod. **253**, 119942 (2020)

15. Nikiforova, N., Putikhin, Y., Shlychkov, D., Frolova, V.: Analysis of Ecosystem Business in Strategic Development. EcoSystConfKlgtu 2023: Ecosystems Without Borders **1**, 16–24 (2023)
16. Kirillova, E.A., Dli, M.I.: Prospects for the formation of innovative ecosystems in industry. Actual Probl. Econ. Manag. **2**(34), 80–94 (2022)
17. Teece, D.J., Pisano, G., Shuen, A.D.: Capabilities and strategic management. Strateg. Manag. J. **18**(7), 509–533 (1997)
18. Lawson, B., Samson, D.A.: Developing innovation capability in organisations: a dynamic capabilities approach. Int. J. Innov. Manag. **5**(03), 377–400 (2001)
19. Agirre, A.: Building up an innovation region through cooperative experiences. Pôle Sud **31**(2), 71–86 (2009)
20. Chernova, G.V., Kalayda, S.A., Khalin, V.G., Yurkov, A.V.: Matters of economic ecosystem classification. J. Appl. Inf. **1**(91), 69–82 (2021)
21. Foray, D., Eichler, M., Keller, M.: Smart specialization strategies—insights gained from a unique European policy experiment on innovation and industrial policy design. Rev. Evol. Polit. Econ. **2**, 83–103 (2020)
22. Foray, D.: On sector-non-neutral innovation policy: towards new design principles. J. Evol. Econ. **29**, 1379–1397 (2019)

Improving the Competitiveness of Russian Industry: Relying on Internal Resources

Tolstobokov Oleg(✉)

Moscow Technical University of Communications and Informatics, Moscow, Russia
tolstobokov@yandex.ru

Abstract. The paper reveals the specific nature of competitiveness improvement of domestic industry based on internal resources employed for development of the national economy. The author emphasizes the need to focus on the requirements for institutions, technological base, and human capital to improve the competitiveness of Russian industry. Russia is bound to restore its competitive positions in the national and global markets, which declined during transition from the old economic system to the new one and from the old industrial base to the advanced one, using technologies appropriate for the new technological structure. Universal neo-industrialization of the Russian economy should be regarded as the imperative of industrial restoration on a new technological base. Increased labor productivity and the system of technical renewal of the national economy are currently among the priorities of its development. This study provides data that can be used to elaborate a strategy for micro and macro managing the promotion of the competitiveness of Russian industry. The author employed instrumental and methodological research apparatus based on principles of systemic and functional analysis.

Keywords: Russian Industry · Competitiveness · Innovation · Internal Resources · Neo-Industrializationintroduction

The economies of different countries of the present day are undergoing profound technological, institutional, and organizational transformations related to transition to a new technological and world economic structure; institutional mechanisms are being transformed, systemic connections are being weakened and undermined, which entails the need for systemic subjects to enter these processes, mutually integrate, and adapt.

The concept of competitiveness improvement of the national economy and its structural links implies a complexly organized two-way process of changes. It is oriented towards the needs of performance and development of the system, which implies both transformation of the system object from the industry to the national economy and transformation of the system receiving it from the regional economy to the global economy, where it is expected to function [1].

Why is this concept needed? It is required to ensure productive interaction of complex systems in conditions of accelerated transformation. At the same time, uneven development pace poses a certain combination of priorities that underlie the concept of competitiveness. These are man, science and technology.

R. Polyakov (Ed.): EcoSyst 2023, LNNS 1092, pp. 33–38, 2024.
https://doi.org/10.1007/978-3-031-67354-2_4

1 Methodology

The research into the concept of competitiveness is in demand due to a wide scope and high significance of the tasks of neo-industrialization faced by the modern Russian economy, namely its technological base. This is not because the Russian economy has undergone market transformations on its way from a command-administrative economy to the modern one, like many countries of the former socialist space. The current conditions of deep technological shifts and global economic transformations bring market systems to a new level of global leadership. Many countries have gone through market transformations; these are the CIS countries, the Baltic countries, the countries of Eastern Europe, Latin America, Cuba, and China. In the USA, the provisions of the new industrial policy of D. Trump included neo-industrialization of the economy as one of the primary issues. This imposes a wide demand on science and practice to develop methodological and theoretical provisions based on the theory of neo-industrialization and aimed at competitiveness of the national economy and its individual components. This paper employs the provisions of the theory of neo-industrialization to improve the competitiveness of the industrial core of the Russian economy, based on a new technological base with consideration of the mistakes of the 'shock therapy' reforms and widespread application of the recommendations of the monetarist theory to the development of the real economy, regardless of economic practice. The research methodology is based on a system approach and other related concepts [2].

1.1 Scientific, Technological and Innovative Potential of the Russian Economy

Promotion of the competitiveness of Russian industry should be performed based on its own technological base, yet this base was significantly undermined in the 90s of the 20th century. The transfer of foreign institutions and technologies into national economy played a specific role in this process. This determined a purely superstructural nature of the current processes of informatization and computerization in the Russian economy. Over the years, the country was focused on the import of technologies, equipment, norms and standards of modern business processes; however, this focus has changed dramatically in recent years. Russia is actively building up its technical, technological and innovative potential. This is evidenced by data of the global competitiveness index: Russia holds 14th position in terms of its technological level, 51st position in terms of its innovative potential, and 42nd position in the field of digital competitiveness. The leaders in this ranking are Switzerland, Sweden, the USA, and Singapore. Russia shows best results in the development of human capital and science – 26th position out of 132 economies in the world. In terms of the state of institutions, the Russian economy dropped from 89th position in 2022 to 110th position in 2023 [3].

Data on investments in research and development by the world's 2,500 largest companies for 2022 updated the record of the previous years and amounted to 1.25 trillion euros, which is 141 billion euros more than that in 2021. The 10 leading countries in this ranking provided over 93% of global expenditures. China increased investments almost 10-fold over 8 years, surpassed the European Union in this indicator and took the 2nd position in the global ranking, behind the American economy [3].

Global R&D expenditures are concentrated in the 'knowledge economy,' which attracts capital and innovation. The leading position among the innovative industries is held by the pharmaceutical and biotechnology industries, as well as the IT sector, which are ahead of the automotive industry, mechanical engineering, industrial equipment industry, military-industrial complex, chemical industry, and consumer goods production. Russia ranked 9th in this indicator in 2019 (Table 1), R&D expenditures calculated in PPP amounted to $44.5 billion, and as a percentage of GDP this indicator attained 1.04% [4].

Table 1. Leading countries in terms of domestic R&D expenditures calculated in PPP of national currencies for 2019

Country	Internal R&D expenditures			
	calculated with regard to PPP of national currencies		calculated with regard to GDP	
	Ranking	USD Billion	Ranking	%
USA	1	657.5	9	3.07
China	2	525.7	14	2.23
Japan	3	173.3	5	3.24
Germany	4	147.5	8	3.18
The Republic of Korea	5	102.5	2	4.64
France	6	72.8	15	2.19
India	7	58.7	45	0.65
Great Britain	8	56.9	22	1.76
Russia	9	44.5	37	1.04
Taiwan	10	44.0	3	3.5
Italy	11	38.8	26	1.45
Brazil	12	36.3	34	1.16
Canada	13	29.3	24	1.54
Spain	14	24.9	31	1.25

Source: Compiled by the author from https://rusinno.ru/article/ [4]

In terms of the share of R&D expenditures in GDP, the world leader is the Republic of Korea (4.64% of GDP), followed by Taiwan, Japan, Germany and the USA. In Russia, the corporate sector brings about 60% of R&D expenditures. However, the return on investment in scientific development is low; the country's contribution to global high-tech production is 0.3%. [5].

Nevertheless, the overall balance in Russian industry is found to be positive despite geopolitical trends and withdrawal of foreign capital that supplied technology.

1.2 Institutional and Socio-Economic Factors of Competitiveness

The concept of competitiveness of the Russian economy requires careful consideration of the requirements for institutions, since the ranking indicators of the Russian economy feature its weak points. The world is moving towards a new institutional structure, and the ways of regulating economic relations are being updated. Centuries ago, the entire economic life, including contracts, legal laws, etc., could be transferred from generation to generation; at present, it requires several cycles of updating institutions and ways of organizing the economic relations. Nowadays, legal norms regulating markets, forms of contracts, economic behavior, and statuses annually change [6].

Development and implementation of the concept of competitiveness needs to be managed using the capabilities of modern management, including strategic management considered within the framework of the reproduction approach [7, 8].

With regard to the above provisions, Fig. 1 presents the requirements for the concept of competitiveness of Russian industry in the national economy:

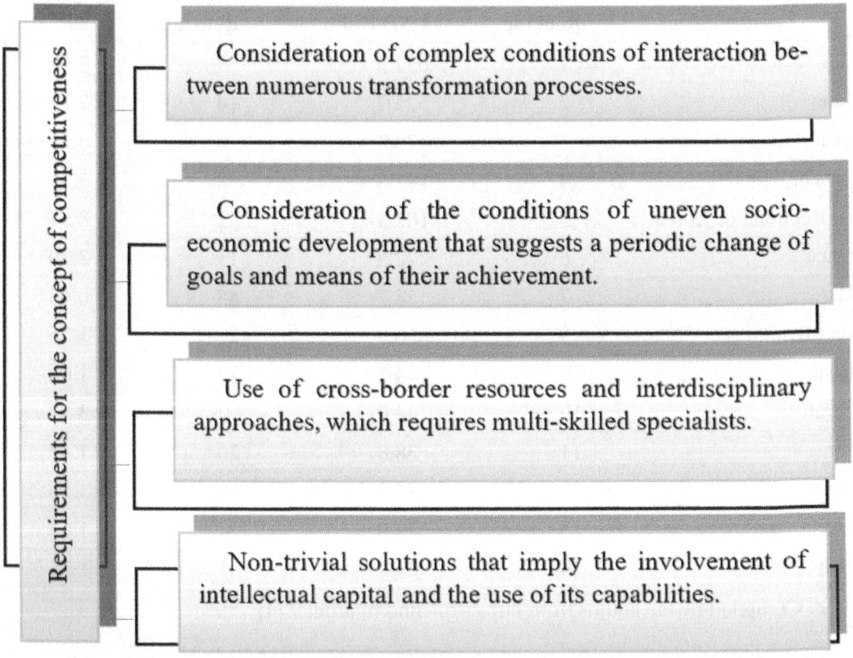

Fig. 1. System requirements for the concept of competitiveness of Russian industry. Source: Compiled by the author based on research materials

The creation of a socially oriented model of spatial development in the Russian economy is relevant and increases the urgency of developing appropriate tools to enhance the competitiveness of Russian industrial enterprises – special contracts (strategic, specific and partnership) [9].

Consider the aspect of effective interaction between enterprises and regional farms, which are complex and integrated. The interaction of complex systems entails gradual convergence of their interests, principles, and decision-making mechanisms between economic entities. Otherwise, there can be asymmetry of cooperation, excessive monopolization of markets, gaps in resource provision, and dominance of the interests of the corporation over the interests of the territories. We assume that one of the strategic management tools to increase the competitiveness is a strategic contract concluded by territories and corporations [10].

Specific contracts can be used to adjust and adapt relationships, which require a long-term period of mutual adaptation and integration in order to work out the contract. In addition, long-term relationships imply confidence and enable insurance and development of detailed and subtle contracts that accurately protect the interests of all parties involved.

Partnership contracts are based on the concept of intellectual capital, which is often considered as a supplement to the regular workforce, but this is another form of economic relations that necessitates a special format of contracts. In recent years, Russia has been integrating into the world economy in a different way, new interactions and cooperation are underway, previous forms of contracts recede into the past and are being relegated to peripheral positions. In these conditions, intellectual capital is of particular demand, since it can yield highly productive capital combinations to meet new challenges in today's world and new requirements of technological development. The demand for intellectual capital and its generation is gradually increasing despite the predominance of corporate capital in the domestic economy, which is engaged primarily in extraction of raw materials and fuel and energy resources. Russia ranks 6th in the Top 10 global ranking for the number of researchers [11].

2 Result and Discussion

For Russian industry, the concept of competitiveness promotion should include the following goals:

- crisis resolution and sustainable economic growth through effective crisis management, management competence, planning, systemic neo-industrialization and development of relevant institutions [12];
- profound qualitative technological and institutional changes that include elaboration of innovative technological developments and digital industry, long-term investments, competent personnel, reliable information resources, advanced centers for research and technology transfer, technology parks, revitalization and containment institutions;
- creation of a comprehensive base of researchers and competent specialists in the field of industrial production based on interdisciplinary approaches, integrative resources, matrix management, and creative (venture) groups [13];
- involvement of intellectual capital to solve complex production problems based on knowledge, intelligence, project management, technologies used to solve research tasks;
- systematic setting of strategic goals and adaptation of industrial policy to changing goals [14].

3 Conclusion

Overall, the temporal imperative of economy integration should be taken into account to improve the competitiveness of Russian industry based on its integration into global and national economic development trends. For integration, it is important to focus on process timing. Integration gravitates towards the present time, towards reality. Integration into the economic system with regard to past values, postulates, ideas, and institutions will definitely be a failure. A look further into the future with its vague images and forecasts applied to current events will bring similar results. The only way of successful integration is to orient towards the existing realities. These are the current day realities, and national economy needs to fit and integrate into them.

References

1. Brizhak, O., Adamenko, A.: Neoindustrial paradigm of Russian corporations integration into the economy system under the conditions of transition to the sixth technological order. Lect. Notes Netw. Syst. **111**, 619–626 (2020)
2. Brizhak, O.V., Tolstobokov, O.N.: Forming the new industrial core of Russian industry: problems and perspectives. Res. Econ. Anthropol. **42**, 155–160 (2022)
3. Global Competitiveness Index for 2022. https://www.weforum.org/ World Economic Forum website
4. Innovative development of Russia (2023). https://rusinno.ru/article/
5. Amy report. UNESCO Institute for Statistics (Montreal, Canada). http://www.unesco.org/new/ru/media-services/single-view
6. Brizhak, O., Polyakov, R.: Creative impetus for the development of innovation clusters of the Russian economy. Lect. Notes Netw. Syst. **474**, 99–104 (2022)
7. Shinkevich, A.I., Zaraychenko, I.A., Ivanova, R.P., Vodolazhskaya, E.L.: Decision making support for the development of new products based on big data technology. IOP Conference Series: Materials Science and Engineering. Krasnoyarsk Science and Technology City Hall of the Russian Union of Scientific and Engineering Associations, pp. 42014 (2020)
8. Seifi, A., Razmkhah, N., Pletnev, D.: The economic dimension of the rights to sustainable development: good governance and human security. E3S Web of Conferences. Ural Environmental Scientific Forum 'Sustainable Development of Industrial Region', UESF 2021 (2021)
9. Kleiner, G.: Strategy for the system harmonization of Russia's economy. Econ. Strat. **5–6**, 72 (2018)
10. Chikanova, E.S., Brizhak, O.V.: Determinants of planning the development of modern economic theory. Russ. Econ. J. **6**, 4–16 (2023)
11. Yermolenko, A., Brizhak, O., Digtiar, O.: New forms of relations arising in the capital flow Advances in Research on Russian Business and Management, pp. 299–317 (2019)
12. Gubanov, S.S.: Powerful breakthrough. Neoindustrialization of Russia and vertical integration. Moscow: *Knizhniy mir* (2012)
13. Buzgalin, A., Kolganov, A.: Critical political economy: the market-centered model of economic theory should be left in the past – notes from the post-soviet school of critical Marxism. Cambridge Journal of Economics, pp. 575 (2015)
14. Chikanova, E.S., Brizhak, O.V.: Determinants of planning the development of modern economic theory. Russ. Econ. J. **6**, 4–16 (2023).

Trends in the Development of the Potential of Ecosystems of the Sharing Economy in the Production Sector

Anastasia Glebova(✉), Anna Khryseva, Irina Ezangina, Elena Abramenko, and Galina Mershieva

Volgograd State Technical University, Volgograd, Russia
nastyag55555@yandex.ru

Abstract. The article is devoted to providing relevant research information on trends in the development of the potential of ecosystems of the sharing economy in the production sphere due to the fact that Russia is currently in conditions of limited access to technology and equipment, and there is a need to maximize the use of its own resources and increase domestic production with the use of domestic element base. At the same time, enterprises are faced with low productivity, irrational use of resources, low efficiency of production facilities, which requires a positive impact of digital transformation. Based on this, the paper highlights the level of development of digitalization of Russian enterprises at the current stage, identifies specific trends in the development of the potential of sharing economy ecosystems in the production sphere, and provides an example in the form of a production capacity sharing platform.

Keywords: Ecosystems · Sharing Economy · Sharing Platform · Production · Digitalization

1 Introduction

Ever since enterprises have been faced with low productivity, poor resource utilization, low capacity efficiency, they need the positive impact of digital transformation.

This is confirmed by Vladimir Gamza, a member of the board of the Russian Chamber of Commerce and Industry, who supported the discussion at the conference "Industry 2023: Digital Technologies". According to the speaker, Russia needs a comprehensive digitalization program. The expert cited data from Rosstat polls conducted among industrialists. About the need for reconstruction of enterprises said 71% of respondents, automation - 55%, the issues of digitalization and new technologies are interested in 35% of respondents. Such data, Gamza stressed, once again confirms that the topic of digitalization is in demand, however, it should be dealt with in the same way as "national programs" [1].

With rapid technological advances, the bigger challenge is how to effectively integrate and coordinate the explosive growth of data resources and even build a multi-tiered intelligent enterprise organization.

R. Polyakov (Ed.): EcoSyst 2023, LNNS 1092, pp. 39–46, 2024.
https://doi.org/10.1007/978-3-031-67354-2_5

And faced with a wave of revolutionary innovative behavior, as well as the growing trend of business models such as the sharing economy, industrial companies began to adapt their traditional state of production to this reality.

The aim is therefore to highlight specific trends in the potential of sharing economy ecosystems in the manufacturing sector. And with that, address the following challenges:

1. To explore the expanding of the boundaries of digital technologies impact on the manufacturing industry;
2. assess the level of development of digitalization of Russian industry;
3. determine the benefits of the sharing economy on production and the level of development ecosystems of the sharing economy;
4. provide examples of Russian and global experience of ecosystems of the sharing economy.

2 Research Materials

When considering the stages of digital transformation, we should focus on understanding the key characteristics in order to advance the process in an orderly manner.

The digital transformation of manufacturing companies is best visualized in 3 stages, characterized in the following table (Table 1).

This table shows that the finish line at this time is the digitization of ecosystems, i.e. the integration stage. Let consider this period.

The creation of an ecosystem includes not only the horizontal connection of enterprises from different industries, but also the integration of upstream and downstream industry chains, so it can only be called a target state for enterprises at this stage.

Ecosystems have their own building blocks, including complementary assets and innovations in distribution among actors, positive externalities, and criteria for calibrating actors with autonomous thinking. These elements enable an ecosystem to generate ecosystem-specific benefits (ESA). Thus, the presence of ecosystem benefits is the main criterion for assessing whether a company is successfully going through the period of integration [2].

The very concept of "ecosystem" is related to the analogue of a natural ecosystem, in which, as in nature, living organisms and their habitat are united, harmonized and balanced, so we focus on the topic of collaboration, which can be defined as a new interactive form in the global environment of the knowledge economy to produce product and process innovations.

Cooperation in ecosystems is not limited to territories or sectors, but brings together different actors on the principles of self-organization and self-management, mutual trust, sustainable development and open innovation. The mechanism of ecosystem self-government through actor coordination has been described in detail by Erkko Autio [3]. In ecosystems, knowledge and information are exchanged, collective learning takes place, and a system of interconnected technologies develops, which contributes to the development of each participant and society as a whole [4].

Shaping this kind of ecosystem requires depth and scale of analysis, especially in contexts of limited contractual relationships and trust, with a gradual transition from one stage of development to the next.

Table 1. Stage characteristics of traditional manufacturing enterprises in the process of realizing digital transformation

Period name	Brief characterization
1. Pilot period (digitization of local terminals)	- Experimental field period of digital transformation of enterprises, which takes place in the early stage of digital transformation - The pilot period is mainly driven by technological progress and the external environment, which lays a solid foundation for the subsequent phases - The aim of the enterprise organization at this stage is to achieve benefits. They hope to reduce their sourcing costs, copying costs, transport costs, tracking costs and verification costs by developing digital technologies so that they can respond to changes in market demand quickly and efficiently - The key resistance level during the pilot period has a different emphasis from a structural point of view
2. Expansion period (digitization of platforms)	- This phase focuses on how to integrate these data resources into various relationships - The platform connects the supply side and the demand side, and at the same time provides an opportunity for both sides to come closer together, thereby reducing communication and coordination costs - In addition to technological advances and the external environment, actors are also the driving force behind the expansion period - It plays an important regulatory role in the process of organizational transformation - Since building a platform requires links between different departments and links, this poses even greater challenges for actors, especially organizers
3. Integration period (digitalization of ecosystems)	- Ecosystems place higher demands on the organization as a whole, as it places greater emphasis on the integration and orchestration capabilities of the enterprise - The integration stage should be platform-based and by linking multiple platforms and multiple facilities into a value ring to realize the joint creation and use of resources between value-added platforms - Through collisions and iterations between platforms, an ecosystem is formed that allows companies to communicate with each other across the industry chain. It is the ultimate goal of companies to achieve digital transformation, that is, to achieve industry-wide integration - Creating a value chain is the main manifestation of the digitalization of the ecosystem

Having understood what an enterprise should strive for and what stages of digitalization it passing through on its journey, let's now look at the importance of the sharing economy for manufacturing.

It is worth noting right away how the sharing business model itself is defined. It can be seen as a structural template that provides a platform for suppliers and customers to customize new ways of creating value with under-utilized (idle) resources. In the sharing economy, practices are carried out on platforms that connect suppliers and customers. By doing so, a potential unused or idle supplier resource of the supplier is turned into a concrete benefit for the customer [5].

This trend as "sharing economy" gives rise to a new way of socialising production such as "co-production", and can expand both the degree and scale of resource sharing through P2P cooperation, thereby enhancing the competitiveness of production. This expands the extent and depth of resource sharing and ensures the integration of production at the societal level.

Traditional production paradigms usually develop with a focus on one area (e.g., technology, society or economics). Compared to these, co-production is more interdisciplinary, revealing the coordinated development of all areas (Fig. 1). Consequently, it is more complex and requires attention from both scientific and sociological perspectives.

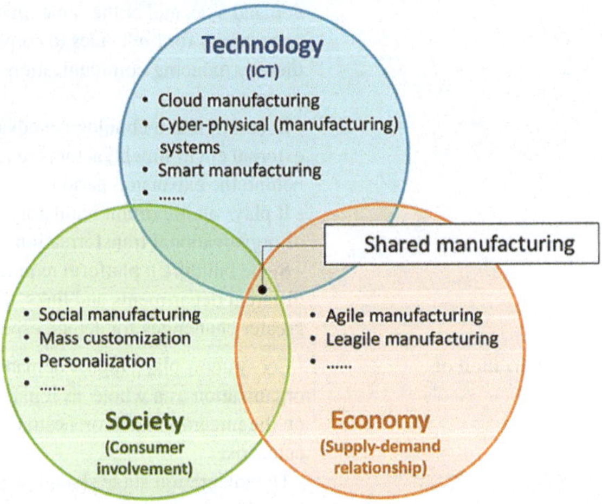

Fig. 1. Examples of production paradigms emphasizing technology, society and economy [6].

And some initial industrial applications have already emerged. For example, Floow2, a Dutch company, is creating an online community where manufacturers can share their idle assets (e.g. equipment, employees and jobs) to generate additional revenue. Haier Group, a new consumer electronics manufacturer in China, plans to create a resource-sharing ecosystem in which all resources involved in all of the company's supply chains are independently available to external partners [6].

The co-production model itself, as a kind of new production node within a social production network, extends the nature of co-production in three forms, which are related respectively to the division of production orders, production resources and production possibilities. Its working logic relies on a platform that brings together common productions in the context of the social production paradigm and is the interface of the production community. This gives us a new way to using generalized productive resources under the guidance of the sharing economy. Both producers and customers, who can simultaneously fulfil different roles in the context of managing commons production, will share the benefits in a "win-win" mode. Any traditional industries, including SMEs, can use this co-production to expand the market, optimize resource allocation, strengthen bargaining power, reduce production costs and increase production flexibility, and inspire creative vitality.

Let us describe the advantages of this model in economic terms. Sometimes factories have sufficient inputs and production capacity but few production orders, while others have sufficient production orders but lack inputs and production capacity. The emergence of joint factories is designed to resolve these kinds of contradictions.

It is best to consider this in the following context. For example, when Covid-19 caused major fluctuations in demand and supply in many sectors: while some companies were left without production capacity, others had to react quickly to sudden spikes in demand.

Enterprises that have their own platforms and are thus in the second stage of digitalization, i.e. the expansion period, can also cooperate with production platforms of the sharing economy, or incorporate elements of their strategy into their own [6].

The list of sectors that could benefit from the B2B sharing economy is long. "This is an opportunity for industries from manufacturing, food and agriculture to energy, mining, chemicals and other sectors," the World Economic Forum said [7]. Capital-intensive, asset-heavy industries with periods of underutilization or redundancy in their ecosystems may have the greatest potential.

The sharing economy is not only good for business: it can also help create a more sustainable global economic model and reduce industrial carbon emissions.

Let's describe the three types of co-production regimes that we mentioned above. These can be used to create appropriate co-production ecosystems: production order sharing, input sharing and output sharing.

For a better explanation, it's worth looking at the classification depending on what it's used for:

1. The first type is used to separate production orders from others. For example, there may be too many production orders to be delivered on time or tailored to specific needs, so these manufacturers or customers will outsource these production orders to general production to fill the gap of insufficient capacity.
2. The second type is used to share own production resources with others. This type of factory may own a lot amount of production resources but does not take production orders to share its own production resources with other producers or customers who lack production resources to expand production.
3. The third type is used to share own production capacity as a service to other producers or customers, which can similarly be considered as a PSS mode. PSS is rather an operation and maintenance mode at the production system level [8].

Figure 2 then provides the most illustrative example of an overview of sharing relationships in production. By looking at the relationships, it is possible to understand changes in production logic and the way benefits are shared, and to elucidate the value creation mechanism associated with such a shared production.

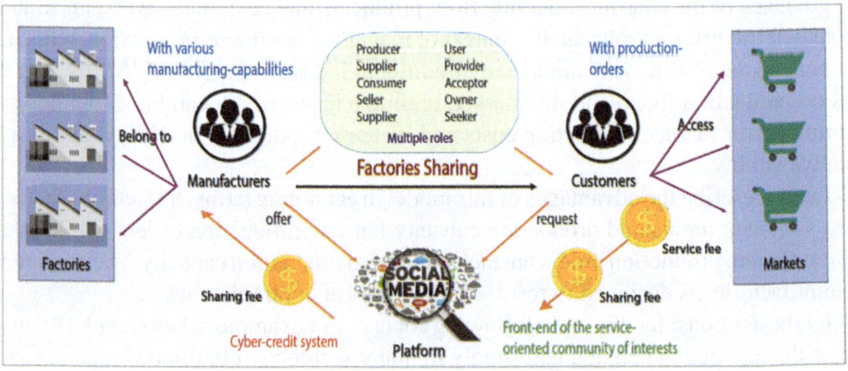

Fig. 2. Overview of sharing relationships in manufacturing [8].

Next, let's look at how Russian enterprises are trying to implement shearing technologies and develop their ecosystems. At the moment, many who are planning to do this have projects at the idea and implementation stage, but it will be very useful to review them in order to see the trend and possibilities of ways to develop this process.

Kirill Lipa, CEO of Transmashholding (TMH), announced an event in 2022, the idea of which is to launch a service enabled heavy and light engineering companies to find equipped production sites online to fulfil orders based on their own drawings. This involves the manufacture of components, tooling or other products that, for one reason or another, cannot be manufactured in an acceptable timeframe in-house. The creation of such a platform will help ensure the fullest possible utilization of capacities (primarily rare and unique equipment) of the largest industrial enterprises in Russia and promptly find contractors for urgent orders in the machine building industry [9].

Now its pilot version is being developed by "TMH Technologies", "2050.integrator" and "Zyfra" [10]. A working group on the project has also been set up on the basis of the Machine Builders Union. The capacity sharing platform should enable the joint utilization of existing industrial plants and their maintenance, and in the near future also supply them with the necessary consumables and components.

Janis Staks, CEO of "2050.integrator", says that the capacity sharing platform will help build co-operation chains and relationships between the customer and the supplier. It will help you quickly find a relevant contractor that is guaranteed to fulfill the order. For example, a company regularly receives additional orders from its customers, which greatly affects its production schedule and leads to changes in logistics chains. At the same time, it is interested in fulfilling the order and increasing production in a moment, even if it lacks its own capacity. This is where SMP (sharing manufacturing platforms) comes in. Potential users of the platform can be all machine-building enterprises in Russia and the EAEU. According to preliminary estimates, in 2025 over 250 enterprises may be connected to the SMP. Attraction of new participants will be carried out by the company "2050.integrator" as the platform develops [11].

In Russia, there are a number of electronic platforms on which orders are placed for metalworking and the manufacture of components and parts. But they operate in the format of noticeboards, the main disadvantage of which is that order fulfillment is not guaranteed. Against this background, SMP claims to the first project of its kind not only in Russia, but also in the world.

Within Russian corporations there are examples of the implementation of information systems for order search. TMH has a Private Industrial Information System (PISP) in operation and is improving it with the support of the Russian Ministry of Industry and Trade. It may become the basis for a future global capacity sharing system. Within the framework of this system, the Contractor service is already working in a similar way, but only within TMH and only for cutting production. The already implemented functionality has the potential for development into a cross-corporate capacity sharing platform [9].

In global practice, there are more examples of how sharing can change industries. Here are some of them.

There are sharing economy manufacturing platforms such as Xometry and Fictiv. They allow businesses to quickly "rent" specialized facilities, such as 3D printing workshops, that would otherwise take significant time and money to set up. Periods of use can be flexible - it can be a one-off period of a few days or a set period once a quarter. For one, this is a convenient way to reduce capital expenditure; for another, it can provide valuable new revenue streams and keep factories and workers busy.

Many other examples are now emerging., Cohealo, for example, allows healthcare providers to share equipment. Platforms such as UK-based Stowga or US firm Flexe help companies that need storage space with businesses that have empty warehouse space to rent. A similar service is available for retail spaces: Need a pop-up showroom for a week to test a new range? Services like Appear Here or Storefront can help you find a suitable property quickly. Intangible assets can also play a role. Through platforms like NineSigma, companies can capitalize on intellectual property such as patents and copyrights, sharing it with innovation-hungry customers [11].

3 Conclusion

The list of sectors that could benefit from the B2B sharing economy is long. "This is an opportunity for industries ranging from manufacturing, food and agriculture to energy, mining, chemicals and other sectors," the World Economic Forum said [7]. Capital-intensive, asset-heavy industries with periods of underutilization or redundancy in their ecosystems may have the greatest potential.

The practices of the sharing economy allow enterprises to harness the power of external resources, which can completely change the business environment. Their use allows sharing economy platforms to realize a diversification strategy that can lead to the embrace of traditional business models platforms [12].

Thus, after investigating the development of the potential of sharing economy ecosystems in the manufacturing sector, we can identify the following trend: due to the complexity of the manufacturing industry chain and the weak technical base of the manufacturing industry in our country, the development of sharing economy in the manufacturing sector is still in its infancy. At the same time, the sharing model can form an industrial ecosystem of sharing, in which various actors such as upstream and downstream enterprises in the industry chain as well as large and medium-sized enterprises can flexibly participate.

The sharing economy is not only good for business: it can also help create a more sustainable global economic model and reduce industrial carbon emissions.

It is clear that this model also demonstrates contributions to production transformation and social innovation. However, there are still some challenges to be addressed, such as how to use new IT technologies to create a more multifunctional platform for integrating different shared factories, how to develop a more rational profit-sharing model, and so on.

Understanding these concepts and their interrelationships can help authorities and organizations make informed decisions, identify opportunities and risks, and navigate the commercial sphere.

References

1. Industry: results of digitalization in 2022 and forecasts — CDO2DAY. https://cdo2day.ru/analytics/promyshlennost-itogi-cifrovizacii-v-2022-godu-i-prognozy/. Accessed 10 Nov 2023
2. Wu, C., Zhang, K., Chen, X.: Research on the digital transformation path of traditional manufacturing enterprises--a three-stage evolutionary model based on structure and actor perspective. J. Shandong Univ. (Philosophy and Social Science Edition) (04), 121–135 (2022). https://doi.org/10.19836/j.cnki.37-1100/c.2022.04.012
3. Autio, E.: Orchestrating ecosystems: a multi-layered framework. Innovation **1**(24), 96–109 (2022)
4. The role of co-operation in the development of integration of industrial enterprises. https://www.mdpi.com/2071-1050/15/9/7180. Accessed 02 Nov 2023
5. Carlborg, P., Snyder, H., Witell, L.: How sustainable is the sharing business model? Toward a conceptual framework. R&D Management (2023)
6. Yu, C.: Shared manufacturing in the sharing economy: concept, definition and service operations. Comput. Ind. Eng. **146**, 106602 (2020)
7. How the sharing economy can transform asset-heavy industries. World Economic Forum. https://www.weforum.org/agenda/2021/05/here-s-how-industries-can-decarbonise-value-chains-while-improving-economics/. Accessed 28 Nov 2023
8. Jiang, P., Li, P.: Shared factory: a new production node for social manufacturing in the context of sharing economy. Proc. Inst. Mech. Eng. Part B: J. Eng. Manuf. **234**(1–2), 285–294 (2019)
9. "Transmashholding" and partners are actively working out a pilot project of machine tool sharing. ROLLINGSTOCK. https://rollingstockworld.ru/proizvodstvo/transmashholding-i-partnery-aktivno-prorabatyvayut-pilotnyj-proekt-sheringa-stankov/. Accessed 07 Dec 2023
10. Zyfra – digitalization technologies to improve the efficiency of industry. GC Zyfra. https://www.zyfra.com/ru/. Accessed 09 Dec 2023
11. The enormous potential of the B2B sharing economy. https://www.hdi.global/infocenter/insights/2022/b2b-sharing-economy/. Accessed 07 Dec 2023
12. Santoso, A.S.: Sharing Economy Practices in Changing Business Environment: Transaction Cost, Resource-Linking, and Hypercompetition Perspectives. Qeios (2023)

Theoretical Aspects of Formation of the Concept of «Banking Ecosystem»

Diana Boboshko[1(✉)], Dmitrii Treistar[1], and Antonina Kulapova[2]

[1] University of Science and Technology MISIS, Moscow, Russia
boboshko.dy@misis.ru
[2] Moscow State Institute of International Relations (MGIMO), Moscow, Russia

Abstract. The article presents the results of a study of the conceptual apparatus used in describing the processes of formation and functioning of banking ecosystems. Advanced search filters were applied to select the most relevant scientific literature, which made it possible to objectively narrow and clarify the scope of bibliometric research. The content of more than 2000 selected sources was analysed, thanks to the technology of clustering key terms. This type of bibliometric research was performed using the VosViewer software tool. The results of the study are presented in the form of a relationship map describing various aspects of digital and banking ecosystems. In course of the analysis of basic definitions of bank-centric ecosystems and practices of their functioning in the Russian Federation, a universal definition of the banking ecosystem was formulated, and its key features were described. The results of this study can be applied both for studying the theoretical foundations of the ecosystem approach to business organization in the relevant field, and in order to form a regulatory framework for regulating the activities of banking ecosystems.

Keywords: Digital Ecosystem · Digital Platform · Banking Ecosystem · Business Ecosystem · Business Model · Vosviewer · Cluster Analysis · Banking

1 Introduction

Currently, business ecosystems in Russia are on the stage of active formation, however, a single definition of the term "digital business ecosystem" has not yet been formed, and the definition of "banking ecosystem" is also missing [1–3]. Digital ecosystems in business remain an unexplored area both in Russian scientific practice and in the world. In recent years, the concept of business ecosystem has appeared more and more frequently in scientific literature and other open sources of information [4–7], but this is not enough to talk about the existence of objective and well-established theoretical foundations for the study of this phenomenon.

The purpose of this research was to study the theoretical foundations of the formation of banking ecosystems and identify their basic characteristics. To achieve this goal, a bibliometric analysis of relevant research on this and other related topics was carried out, as well as a cluster analysis of the interconnections of key terms, and the main characteristics of bank-centric ecosystems were identified.

R. Polyakov (Ed.): EcoSyst 2023, LNNS 1092, pp. 47–56, 2024.
https://doi.org/10.1007/978-3-031-67354-2_6

2 Methods and Materials

Due to the fact that Scopus and Web of Science (WoS) databases of peer-reviewed scientific literature have restricted access to their resources in Russia, the analysis was carried out on the basis of an open platform for researchers and scientific organizations, The Lens [8], since it is the most complete one and even exceeds the leading global commercial databases in its width [9].

To achieve the objectivity in this study, a primary analysis of statistical data on the number of scientific publications in various fields of ecosystem research for the period from 1970 to 2022 was carried out. The results of the statistical analysis are presented in Fig. 1.

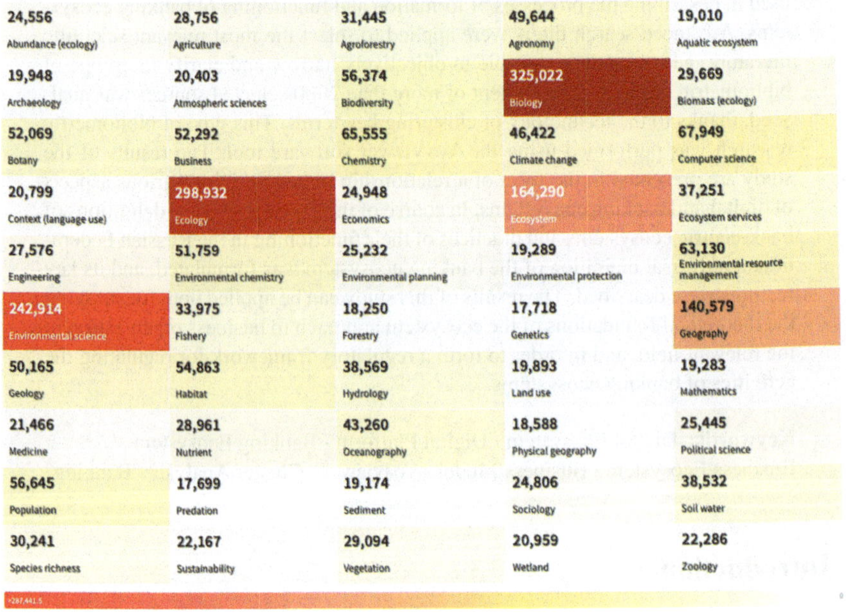

Fig. 1. Statistics on the number of scientific publications in different areas of ecosystem research in The Lens database for the period from 1970 to 2022 [10]

The data from Fig. 1 made it possible to conclude that in the context of the last 50 years, the most significant areas of ecosystem research were natural sciences: biology (325,022), ecology (298,932), environmental science (242,914), geography (140,579) and others. At the same time, the analysis revealed 52,292 publications that meet the theme "Business", and 24,948 works that meet the theme "Economics", yet it could not be confirmed, whether their use is appropriate in the context of studying ecosystem business models. The analysis of the content of these works has established that their predominant majority basically considers the economic aspects of various issues of the development of natural ecosystems, which means that they have no direct relation to business models. Accordingly, further investigation of the nature of banking digital ecosystems is impossible without careful selection of scientific literature.

In order to get the most objective picture of the results of bibliometric analysis of scientific networks, when selecting an array of scientific papers on The Lens platform, the following filters were applied: "publication topic", "publication type", "subject" and "publication year". The request is formed from four different topics (in Table 1 – "publication topic"). The search query filters used in the selection are shown in Table 1.

Table 1. Request filters used in the selection of data set on The Lens platform

№	Filter	Selected values
1	«publication topic» =	(Digital ecosystems, Business ecosystems, Digital banks, Business models)
2	«publication type» =	(Journal article, Dissertation, Conference proceedings article, Book, Report)
3	«subject» =	(Business and International Management, Economics and Econometrics, Management of Technology and Innovation, Information Systems, General Business, Management and Accounting, Software, Marketing, Modelling and Simulation, Business, General Economics, Econometrics and Finance, Finance, Information Systems and Management)
4	«publication year» =	(2002, 2003, 2004, 2005, 2006, 2007, 2008, 2009, 2010, 2011, 2012, 2013, 2014, 2015, 2016, 2017, 2018, 2019, 2020, 2021, 2022, 2023)

As a result of the analysis, 1,690 scientific papers that meet the search criteria described above were selected, of which 1,436 were published after 2015. Statistics on the number of publications by year in the period from 2002 to 2022 are shown in Fig. 2.

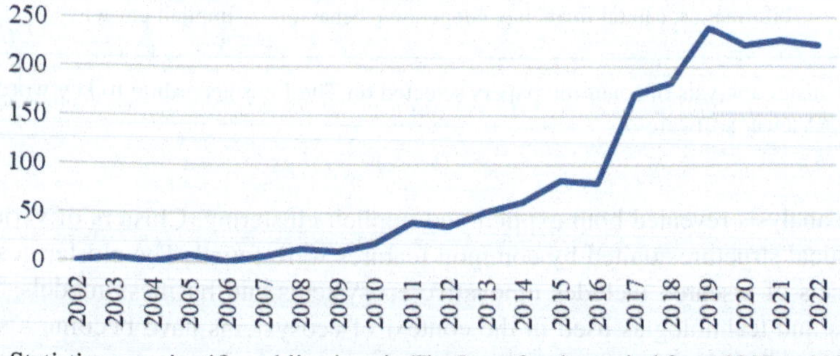

Fig. 2. Statistics on scientific publications in The Lens for the period from 2002 to 2022. [10]

The figure shows the trend of growth in the number of scientific papers which meet the request, based on the filters from Table 1. The general trend demonstrates the growing popularity of the topic of digital business ecosystems on The Lens platform, which once again confirms the relevance of this study. Undoubtedly, such a trend is directly related to the development of information technologies and their impact on the economy.

For a more objective and detailed assessment of the content of the scientific papers previously selected on The Lens platform, a cluster analysis of the connections between key terms ("key words") was carried out. It is based on a bibliometric study and is performed using the VosViewer software tool. Cluster analysis made it possible to build a map of current research areas concerning the concepts of "Digital ecosystems", "Business ecosystems" and "Business models", as well as to identify various semantic and structural links between them and other areas of research. The results of cluster analysis in the form of a relationship map and conclusions on it are presented in Fig. 3.

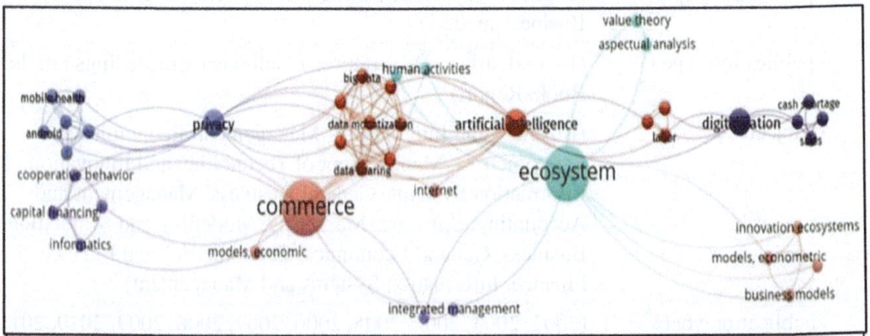

No.	Cluster	Conclusion on cluster
1	Innovation ecosystems, models, business models	Describes the research object
2	Big data, privacy regulation, data monetization, data sharing, digital technology, social media, internet of things, data sharing, artificial intelligence, structuration theory	Describes various research methods and technologies
3	Sales, employment, entrepreneurship	Describes functions
4	Informatics, capital financing, cooperative behavior	Implicit cluster

Fig. 3. Cluster analysis of scientific papers selected on The Lens according to key word carried out by VosViewer software

The analysis revealed both explicit and implicit clustering. Clusters of various terminological structures united by common features were found. The cluster describing the objects of research includes innovative ecosystems and business models. Various methods and technologies used in the context of ecosystems have become a separate cluster as well. Other significant connecting elements, which were found, are "ecosystem", "commerce" and "privacy". The connecting elements identified during the analysis made it possible to assess how the formed clusters are interconnected, as well as to draw conclusions about the nature of their interdependence.

3 Main Results

A comprehensive analysis of the obtained data made it possible to form the most objective view of digital ecosystems and their features. The interpretation of the results of cluster analysis in terms of the identified connecting elements is shown in Fig. 4.

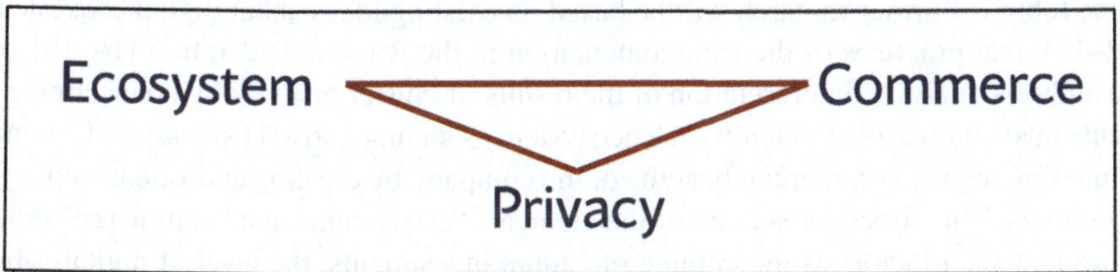

Connecting elements	Identified using the map	Interpretation
Ecosystem	Human activities, commerce, value theory, integrated management, digitalization, innovative ecosystems, business-models, artificial intelligence	Describes the ecosystem as an innovative business model, notes the connection of the ecosystem with the theory of value creation (for the consumer) and an integrated management system. Digitalization is the main driver.
Commerce	Information technologies, capital investments, economic models, ecosystems, Internet, integrated management	Describes commerce as an important area of case studies. Connects ecosystem models with commercial benefits and capital investments. The importance of IT
Privacy	Mobile applications, e-health, human activities, data monetization, data exchange, big data	Describes the connection between the ecosystem and the issue of privacy. There is a connection between human activities, mobile applications, data exchange and their confidentiality. The issue of data monetization is noted.

Fig. 4. The interpretation of the results of cluster analysis in terms of the connecting elements

Interpretation of the results of cluster analysis by connecting elements made it possible to identify the ecosystem as an innovative business model aimed at maximizing the commercial benefits of the company by creating additional values for consumers. The direct connection of the elements "ecosystem" and "commerce" establishes that a special role in the formation of the business ecosystem and the maintenance of its work is played by such factors as the volume of capital investments, the level of digitalization and the degree of introduction of information technologies. The direct connection of the elements "commerce" and "confidentiality" highlights the problem of commercialization of personal data of users of the ecosystem, the possibility of their use for the purposes of the company.

The analysis of the scientific literature made it possible to go deeper into the field of ecosystem development and conclude that the theoretical foundations for studying this phenomenon are still actively being formed. It is worth noting that the topic of digital ecosystems is also quite popular in the Russian scientific community: 444 scientific papers were found (as of 2023/04/08) in the largest Russian scientific library eLibrary [11] at the request of "Banking ecosystems" with partial application of filters from Table 1. Further research will be based on existing ideas about digital ecosystems [12–15], real practices of their implementation in the Russian Federation [16–19] and the author's analysis. Interpretation of the results of cluster analysis by connecting elements made it possible to identify the ecosystem as an innovative business model aimed at maximizing the commercial benefits of the company by creating additional values for consumers. The direct connection of the elements "ecosystem" and "commerce" establishes that such factors as the volume of capital investments, the level of digitalization and the degree of introduction of information technologies by the company play a special role in the formation of the business ecosystem and the maintenance of its work [20, 21]. The direct connection between the elements "commerce" and "privacy" highlights the problem of commercialization of personal data of an ecosystem users, and the possibility of its use for the purposes of the company.

For the first time, the concept of business ecosystems was developed and presented in the 1990s by the American entrepreneur James F. Moore, who borrowed the term "ecosystems" from another science – biology. He gave the following definition of an ecosystem: a set of economic entities that serves as the basis for interaction between various members of the business community [22, 23]. Moore found that business ecosystems allow the companies involved to provide comprehensive solutions that include a "complete package of values" for the client, which no single company could offer on its own. Among the elements of the ecosystem the entrepreneur listed suppliers, manufacturers, competitors and all other stakeholders.

The concept formulated by Moore was reformed in the 2000s by an American professor in the field of strategy and entrepreneurship – Adner Ron. In his writings was highlighted the importance of client-centricity [24, 25] and shifting the focus from the manufacturer to the consumer. Adner defined an ecosystem as a coordinated structure of a multilateral set of partners who need to interact in order to materialize a value proposition for the client. Based on his principles, it also became clear that it is not the company that determines the value proposition, but the consumer themselves.

In 2018, the concept of business ecosystems was subjected to new adjustments, but this time by a British economist specializing in digital ecosystems, Michael G. Jacobides [26]. In his works, it is noted that the emergence of ecosystems is actually facilitated by modularity and platforming, since they allow a number of separate but interdependent organizations to coordinate their actions without a complete hierarchical structure. Jacobides significantly shifted the emphasis of business ecosystem research, first towards innovation, and later towards digital technologies.

In Russia, active study of the phenomenon of ecosystems has begun relatively recently. The most complete definition of the term "digital ecosystem" was given by the main emission and monetary regulator of the country in 2021 [27]. The Bank of Russia has established that an ecosystem is a set of services built on the basis of customer data, allowing users to receive a wide range of products and services within a single process. A summary overview of the definitions of business and digital ecosystems is given in Table 2.

Table 2. Definitions of business and digital ecosystems by different author

Term	Definition	Author
Business ecosystem	A set of economic entities that serves as the basis for interaction between various members of the business community [22]	James F. Moore
Business ecosystem	A coordinated structure of a multi-stakeholder set of partners who need to interact in order to materialize a value proposition for the client [24]	Ron Adner
Business ecosystem	A set of interacting organizations created on a modular basis, which are united by the impossibility of redistributing collective investments. Ecosystems are based on innovation and digital technologies [26]	Michael J. Jacobides
Digital ecosystem	A comprehensive project that unites a large number of participants, information services and business processes, based on the principles of mutually beneficial relations ("win-win") [27]	Higher School of Economics
Digital ecosystem	A digital space in which many services of one company or several partner participants function smoothly [28]	ECM–Journal
Digital ecosystem	A set of services based on customer data that allows users to receive a wide range of products and services within a single process [29]	Bank of Russia
Digital ecosystem	A customer-centric model that combines two or more groups of products, services, and information (of its own production and/or other partner) to meet the final needs of customers (security, housing, entertainment, etc.) [30]	Ministry of Economic Development of the Russian Federation

In the classical sense, a business ecosystem is a set of own or partner offers, united around a specific company [31–33]. Therefore, a bank-centric ecosystem is characterized by the fact that a bank is located in the center of such an ecosystem. At the same time, if the key elements of the ecosystem acquire a digital form, and their interconnection is achieved through digital technologies, such an ecosystem will be called digital [34, 35]. This leads to the conclusion that the most accurate definition of a bank–centric digital ecosystem (hereinafter referred to as the banking ecosystem) will be a network of services based on various platform solutions and digital technologies, with a bank serving as a base for the system. The purpose of such an ecosystem is to maximize the commercial benefits of the bank by creating additional values for consumers.

4 Conclusion

Thus, based on the analysis carried out, the following comprehensive definition can be proposed: the banking ecosystem is a network of services on various platform solutions and digital technologies, with a bank at its core. It is necessary to formulate a definition of the banking system not only from a theoretical point of view, which is still important, but also in order to regulate the activities of such ecosystems, as this issue has become especially relevant in recent years.

Based on the conducted research of the theoretical basis for the formation of ecosystems in business, were identified specific features and signs by which a banking ecosystem can be identified. To such key features have been attributed the following:

- the bank acts as the center of the ecosystem, i.e. all its own or partner ecosystem services are formed around its brand and coordinated by it;
- the bank uses advanced digital technologies, which ensure stable operation of all mechanisms of the ecosystem;
- integration of the digital banking platform into a common ecosystem environment;
- a banking ecosystem meets both financial and non-financial needs of customers.

As a result of the study, a generalized definition of the ecosystem type under study was formulated: a banking ecosystem is a network of services of both financial and non-financial nature based on various platform solutions and digital technologies, with a bank at its core. The introduction of a well-established definition of the concept of "banking ecosystem" has not only theoretical significance, but also practical significance for the purposes of regulating the activities of such ecosystems.

References

1. Matkovskaya, Y.S., Vechkinzova, E., Biryukov, V.: Banking ecosystems: identification latent innovation opportunities increasing their long-term competitiveness based on a model the technological increment. J. Open Innovation: Technol. Market Complexity **8**(3), 143–175 (2022)
2. Boboshko, D.Y., Treistar, D.M.: Analysis of the formation and development prospects of banking ecosystems in the Russian federation. Innovative Econ. Dev. **3**(75), 111–121 (2023)
3. Boboshko, D.Y., Treistar, D.M.: Theoretical basis for the formation of ecosystem models of Russian banks. Bull. Siberian State Ind. Univ. **3**(45), 94–103 (2023)
4. Parker, G., Van Alstyne, M., Jiang, X.: Platform ecosystems: how developers invert the firm. MIS Q. **41**(1), 255–266 (2017)

5. Tolstykh, T.O., Alpeeva, E.A., Boboshko, D.Y., et al.: Ecosystem approach for assessing the socio-economic development of industrial and regional systems in the context of digitalization. Cooperation Sustain. Dev. **245**, 1609–1618 (2022)
6. Bogers, M., Sims, J., West, J.: What is an ecosystem? Incorporating 25 years of ecosystem research. Acad. Manag. Proc. **2019**(1), 1–29 (2019)
7. Gamidullaeva, L., Shmeleva, N., Tolstykh, T., Shmatko, A.: An assessment approach to circular business models within an industrial ecosystem for sustainable territorial development. Sustainability **14**(2), 704 (2022)
8. The Lens – Search results. https://link.lens.org/rA8MDfNyRZ. Accessed 07 Apr 2023
9. Volkov, A., Rishko, Y., Boboshko, D., Eliseeva, E., Treistar, D., Kulapova, A.: Multitrack educational programs as a method of educational process personalization at universities. Educ. Sci. **14**(1), 22 (2024)
10. The Lens – search query "Ecosystem". https://clck.ru/35v3Vk. Accessed 07 Apr 2023
11. ELIBRARY. https://www.elibrary.ru/. Accessed 08 Apr 2023
12. Yu, G., Li, W., Shao, M., et al.: Ecosystem science research and ecosystem management. Acta Geogr. Sin. **75**(12), 2620–2635 (2020)
13. Gamidullaeva, L., Tolstykh, T., Bystrov, A., Radaykin, A., Shmeleva, N.: Cross-sectoral digital platform as a tool for innovation ecosystem development. Sustainability **13**(21), 11686 (2021)
14. Fenwick, M., Vermeulen, E.P.M.: Banking and Regulatory Responses to FinTech Revisited–Building the Sustainable Financial Service 'Ecosystems' of Tomorrow. Singapore J. Legal Stud. 165–189 (2020)
15. Albarrak, M.S., Alokley, S.A.: FinTech: ecosystem, opportunities and challenges in Saudi Arabia. Risk Financ. Manag. **14**(10), 460 (2021)
16. Ivanovich, M.N.: Ecosystem of PJSC Sberbank. Ecosystem products for legal entities. Adv. Sci. Educ. **14**, 68 (2020)
17. Yuryev, D.A., Ermolaev, K.N., Nedorezova, E.S.: The banking services market in the innovative economy development. In: Ashmarina, S.I., Mantulenko, V.V. (eds.) Proceedings of the International Conference Engineering Innovations and Sustainable Development, LNCE, vol. 210, pp. 425–432. Springer, Cham (2022). https://doi.org/10.1007/978-3-030-90843-0_49
18. Tolstykh, T., Gamidullaeva, L., Shmeleva, N., Lapygin, Y.: Regional development in Russia: an ecosystem approach to territorial sustainability assessment. Sustainability **12**(16), 6424 (2020)
19. Tolstykh, T., Shmeleva, N., Gamidullaeva, L.: Evaluation of circular and integration potentials of innovation ecosystems for industrial sustainability. Sustainability **12**(11), 4574 (2020)
20. Boboshko, D.Y.: Digitalization in small business tax administration. Ind. Competitiveness: Digitalization Manag. Integr. **1**, 675–687 (2020)
21. Sidorova, E., et al.: Formation of a sustainable mechanism of preferential VAT taxation of exports as evidenced by the Russian federation practice. Economies **9**(4), 190 (2021)
22. Moore, J.F.: The death of competition: leadership & strategy in the age of business ecosystems. Harper Business, USA (1996)
23. Moore, J.F.: Business ecosystems and the view from the firm. Antitrust Bull. **51**(1), 31–75 (2006)
24. Adner, R.: Match your innovation strategy to your innovation ecosystem. Harv. Bus. Rev. **84**(4), 98 (2006)
25. Adner, R.: Sharing value for ecosystem success. MIT Sloan Manag. Rev. **63**(2), 85–90 (2022)
26. Jacobides, M.G., Cennamo, C., Gawer, A.: Towards a theory of ecosystems. Strateg. Manag. J. **39**(8), 2255–2276 (2018)
27. HSE article: Digital ecosystems. https://hsbi.hse.ru/articles/tsifrovye-ekosistemy/. Accessed 13 Apr 2023
28. ECM JOURNAL article: Digital ecosystem. https://ecm-journal.ru/material/cifrovaja_ehkosistema_modnyjj_termin_ili_novaja_realnost. Accessed 13 Apr 2023

29. An official Bank of Russia publication: Regulation of risks of banks' participation in ecosystems and investments in immobilized assets. http://www.cbr.ru/Content/Document/File/123688/Consultation_Paper_23062230.pdf. Accessed 13 Apr 2023
30. An official Ministry of Economic Development of the Russian Federation publication: The concept of general regulation of the activities of groups of companies developing various digital services based on a single ecosystem (In Russ.). https://www.economy.gov.ru/material/file/cb29a7d08290120645a871be41599850/koncepciya-21052021.pdf. Accessed 13 Apr 2023
31. Wieringa, R.J., Engelsman, W., Gordijn, J., Ionita, D.: A business ecosystem architecture modeling framework. In: IEEE 21st Conference on Business Informatics (CBI) 2019. vol. 1, pp. 147–156. IEEE, USA (2019)
32. Nambisan, S., Zahra, S.A., Luo, Y.: Global platforms and ecosystems: implications for international business theories. J. Int. Bus. Stud. **50**, 1464–1486 (2019)
33. Rong, K., Lin, Y., Li, B., Burström, T., Butel, L., Yu, J.: Business ecosystem research agenda: more dynamic, more embedded, and more internationalized. Asian Bus. Manag. **17**, 167–182 (2018)
34. Senyo, P.K., Liu, K., Effah, J.: Digital business ecosystem: literature review and a framework for future research. Int. J. Inf. Manage. **47**, 52–64 (2019)
35. Petrova, M., Popova, P., Popov, V., Shishmanov, K., Marinova, K.: Digital Ecosystem: Nature, Types and Opportunities for Value Creation. In: Rodionov, D., Kudryavtseva, T., Skhvediani, A., Berawi, M.A. (eds.) 3rd International Scientific Conference on Innovations in Digital Economy 2021, CCIS, vol. 1619, pp. 71–85. Springer, Cham (2022). https://doi.org/10.1007/978-3-031-14985-6_5

The Role of Small and Medium-Sized Businesses in the Green Economy

Sergey Bank[1], Viktor Shnaider[2](✉), Dmitry Shlychkov[2], and Vlada Tereshina[1]

[1] Russian Technological University, Moscow, Russia
[2] Financial University Under the Government of the Russian Federation, Moscow, Russia
VVShnajder@fa.ru

Abstract. In the context of contemporary challenges and global climate change, the importance of sustainable societal development is increasing. Sustainable development of society is directly related to the sustainability of its economy. Small and medium-sized enterprises play a significant role in the emergence of a "green" economy and can also become key agents in promoting sustainable development. There are numerous ways in which small businesses can enhance their environmental responsibility, including reducing emissions of pollutants into the atmosphere, waste management and control, and increasing environmental accountability, among others. In this article, the positive aspects of small and medium-sized businesses transitioning to a "green" course are demonstrated, as well as the aspects that play a significant role in their development. The experience of foreign countries shows that the environmental awareness of enterprises has a favorable impact on their growth, as well as on the overall economic development of the country. Government support for small and medium-sized businesses facilitates a rapid transition to a "green" path for businesses and provides them with the opportunity to become drivers of sustainable development and innovation, thereby contributing to a more sustainable future for the country.

Keywords: "Green" Economy · Environmental Awareness · Small Enterprises · Adverse Impact · Business Models · "Green" Course · Ecology · Small And Medium-sized Businesses · Environment

1 Introduction

Environmental awareness in the modern world is one of the key challenges facing humanity. One of the factors that promote this awareness is the adoption of the "green" economy, which contributes to minimizing the negative impact on the environment. The orientation towards a "green" course for small and medium-sized businesses plays a significant role in creating sustainable business models and improving the environmental situation.

The "green" economy is a direction in economic science that considers the economy as an interdependent component of the natural environment within which it exists and is a part of it. It aims to preserve the well-being of society through the efficient use of natural resources and the return of end-use products into the production cycle [1].

R. Polyakov (Ed.): EcoSyst 2023, LNNS 1092, pp. 57–64, 2024.
https://doi.org/10.1007/978-3-031-67354-2_7

The theory of the "green" economy is based on three axioms: it is impossible to indefinitely expand the sphere of influence in a limited space; it is impossible to demand the satisfaction of endlessly growing needs in conditions of limited resources; everything on the Earth's surface is interconnected [2].

Small and medium-sized enterprises have significant potential for implementing innovative processes aimed at reducing the negative impact on the environment. Such processes may include sustainable use of natural resources, the application of efficient waste management methods, and the creation of environmentally friendly products, among others.

The analysis of the role of small and medium-sized businesses in the "green" economy is of paramount importance because it contributes to the development of strategies that ensure a balance between economic interests and environmental needs. The effective implementation of sustainable business models by small and medium-sized enterprises can be a key factor in achieving environmental awareness.

2 Methods and Materials

Environmental awareness is one of the competitive advantages in the business environment. The experience of foreign countries clearly shows that the "green" economy of small enterprises has a positive impact on the overall economy of the country and its environmental footprint. In global practice, the "green" economy and "green" investment are sustainable trends. Representatives of the business community consider the "green" economy as a direction for reforms that allow for the transformation of the economy, even in conditions of reduced funding and project profitability challenges, to enhance its global competitiveness and ensure environmental safety. The "green" economy is also seen as a driver that stimulates innovative demand [3].

Significant contributions to the development of the "green" economy concept have been made by both Russian and foreign scientists, such as S.I. Kodaneva, M.P. Efremova, N.A. Efremov, I.E. Ilyakova, N.I. Ivanova, L.V. Levchenko, and others.

The primary principle of environmental awareness for enterprises is the six "R" rule: reduce, reuse, recycle, replace, refill, repair. The goal of the "green economy" is economic growth, which is possible with both government and private investment in the "green sectors" of the economy and the development of "green skills" among entrepreneurs and businesses, making it highly relevant [4].

The application of motivational methods in small and medium-sized businesses contributes to the creation of sustainable business models, which directly impact the environmental awareness of both businesses and the general population.

Various sources of information were utilized during the writing of this article, including scientific works, articles, internet resources, and more.

3 Discussion

The "green" economy in small enterprises necessitates a change in existing business models. Most companies continue to base their sales model on increasing the quantity of products sold, regardless of their material consumption [4]. Consequently, small enterprises have a negative impact on the environment in pursuit of increased profits from their product output.

Global experience demonstrates the successful transition of countries to a higher level of development, which involves the implementation and utilization of "green" technologies in various economic sectors [5].

By applying the six "R" rule and with government support, it is possible to promote the "green" growth of small and medium-sized enterprises. Government support can encompass the following directions (see Fig. 1).

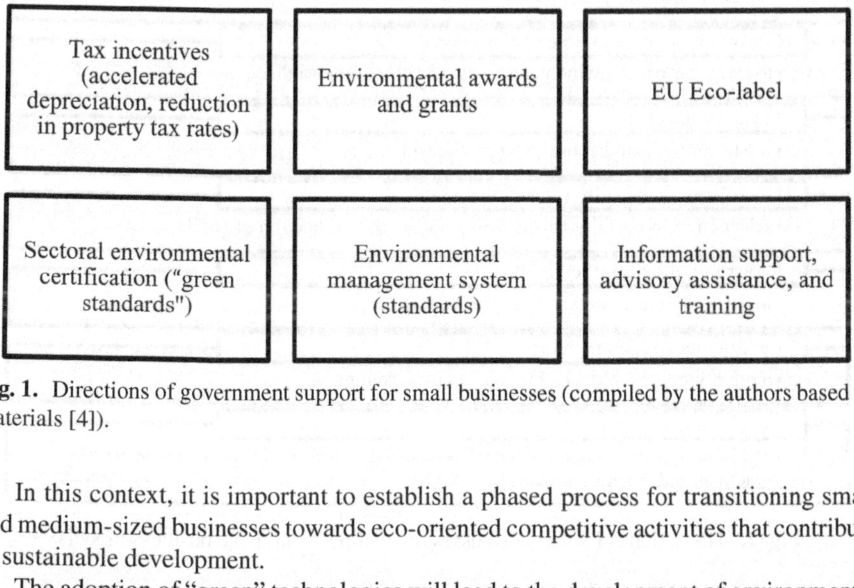

Fig. 1. Directions of government support for small businesses (compiled by the authors based on materials [4]).

In this context, it is important to establish a phased process for transitioning small and medium-sized businesses towards eco-oriented competitive activities that contribute to sustainable development.

The adoption of "green" technologies will lead to the development of environmental entrepreneurship in Russia. The experience of foreign countries attests to the investment attractiveness of eco-business, as it, in conjunction with government support, enables the development and production of innovative, environmentally oriented products [5]. A systematic progression in this direction has already allowed the United States, Japan, and Western European countries to take leading positions in the adoption of environmental technologies and the export of eco-products. Therefore, this development direction is relevant for Russia [6].

4 Research Results

"In practical terms, the implementation of the principles of the 'green' economy entails fundamental changes in three main directions: rational management of water resources (rational consumption and purification); the adoption of circular models (maximizing the lifespan of existing material resources, reconfiguring waste disposal technologies); and energy transition" [7].

It is also important to note the role of consumer demand in the issue of environmental business innovations. "More environmentally-sensitive consumers will raise their expectations regarding the environmental quality of products and services provided by manufacturers. Scientific research has confirmed the importance of consumer demand not only for the adoption of product eco-innovations as a whole but also for the novelty/radicalness of such eco-innovations" [8].

The role of small and medium-sized enterprises in creating sustainable business models can benefit the environment through a series of specific actions (see Fig. 2).

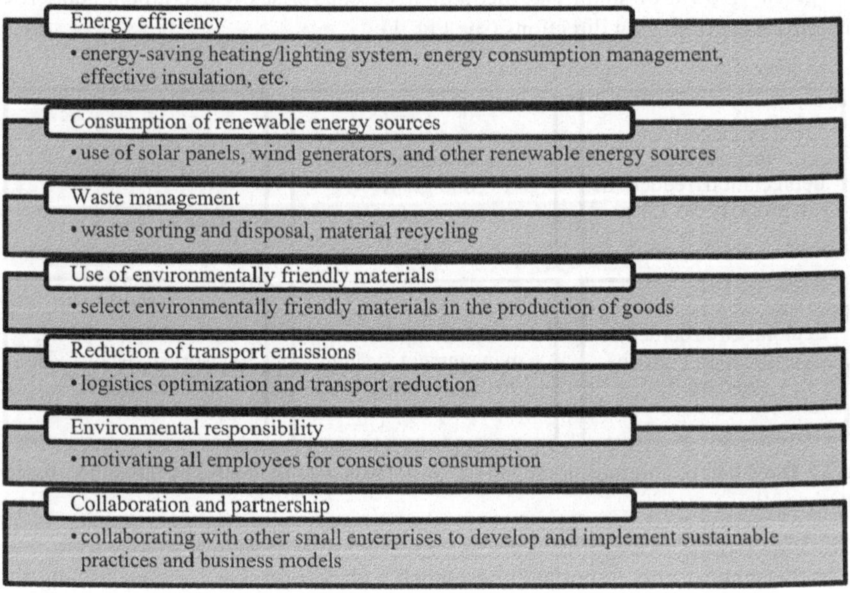

Fig. 2. The actions of small businesses in creating sustainable business models.

With the support of government assistance, small businesses can minimize their negative impact on the environment by ensuring specific procedures (see Fig. 3). These procedures can help minimize the adverse effects, turning them into positive ones.

The path to environmental awareness in small businesses is a fairly labor-intensive and lengthy process. Negative factors require persistent efforts, necessitating more time to transition to a 'green' course. However, the consequences of this transition positively impact not only the competitiveness and profitability of the enterprise but also the country's economy as a whole. Some of the positive outcomes include:

– reduced environmental impact. Enterprises transitioning to the 'green' economy minimize their negative impact on the environment by reducing emissions of pollutants, decreasing water and energy resource consumption, and ensuring rational waste management;

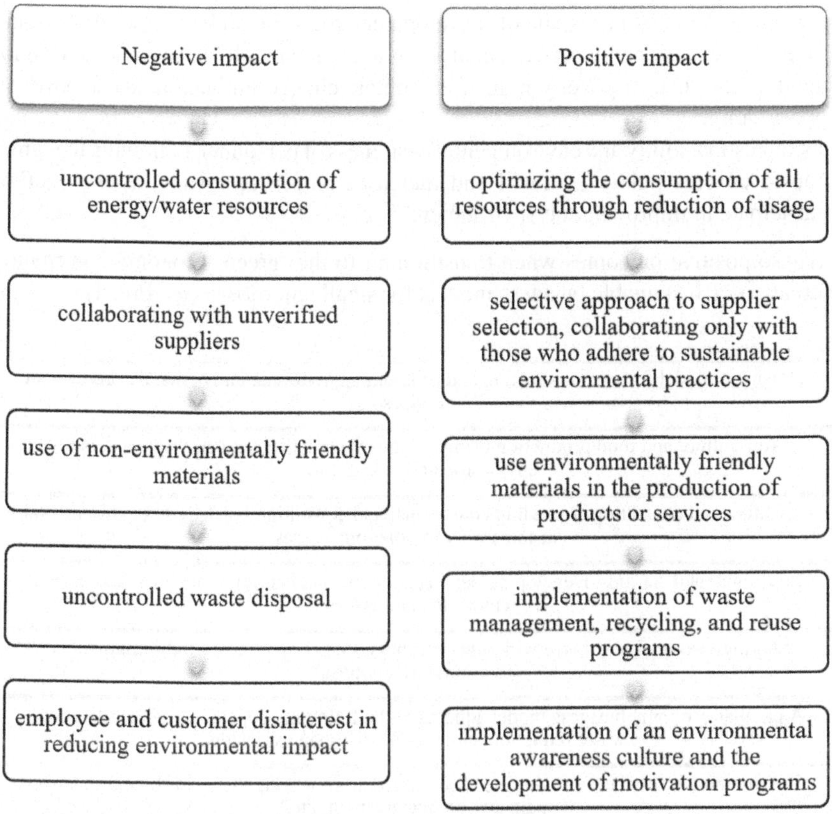

Fig. 3. Procedures to reduce the negative impact on the environment by small businesses.

- economic benefits. Lower energy and raw material expenses lead to reduced operational costs, enhancing the financial stability of the enterprise and increasing profits;
- job creation. Transitioning to renewable energy sources and implementing ecological technologies leads to the creation of new jobs;
- market opportunities. Embracing the 'green' economy enables small businesses to develop and produce eco-friendly products, expanding their customer base and sales;
- risk reduction. Using environmentally friendly resources provides security in the face of fluctuating energy and raw material prices;
- enhanced company image. Establishing a positive company image through the transition to the 'green' economy attracts new environmentally conscious customers and investors;
- competitive advantages. The 'green' course stimulates innovation, such as the adoption of ecological technologies and the use of environmentally friendly production materials;

- contribution to global sustainable development goals. Small businesses transitioning to the 'green' economy actively contribute to achieving global sustainable development goals, such as poverty reduction, climate change mitigation, and biodiversity preservation;
- social responsibility and environmental awareness of personnel. Following the 'green' course ensures public goodwill and increases employee loyalty, as they actively participate in improving the environment.

These positive outcomes when transitioning to the 'green' economy contribute to the creation of sustainable business models for small enterprises (see Fig. 4).

A business model focused on creating and promoting environmentally-friendly products or services

Agriculture and food production using sustainable practices and ensuring access to environmentally clean food

Education and consulting. Providing educational and consulting services on environmental sustainability to other businesses

Environmental startups. Developing new technologies and products aimed at addressing environmental issues

Sharing resources and space with other businesses can reduce costs and minimize the environmental footprint

A circular economy business model adheres to the following concepts: waste utilization, waste reduction, and product lifespan extension

Environmental service. Providing services aimed at improving the environment (energy audits, waste management, etc.)

Sustainable financing. Investors can provide financial support to small businesses with sustainable models

Donations. The practice of donating and participating in social and environmental initiatives to support sustainable development

Fig. 4. Potential sustainable business models for small enterprises.

It is essential to note that when choosing a specific business strategy, it is important to consider factors such as:

- the business direction;
- market conditions;
- the goals of the enterprise.

The application of sustainable business models can serve both as a measure of social and environmental responsibility and as a source of competitive advantages, contributing to the growth and development of small enterprises. "Most countries understand the global trend associated with the energy transition and the need to transform not only their own economy but also existing models of consumption and waste management. The Russian Federation is taking its first steps in this direction, although overall, this movement is somewhat fragmented and unbalanced" [9].

"According to the FiBL research, the Russian Federation is classified as a country in the process of developing green policy legislation... However, it is important to note that the Russian Federation is currently taking stringent measures to ensure environmental safety, intensifying its development in the field of the 'green economy,' and promoting environmental awareness" [10].

The questions of the green economy - the economy of the future, and the directions of digitalization are closely intertwined with the aspects of sustainable development, which are studied by leading Russian and foreign specialists, including the scientific works of O.V. Efimova, E.V. Nikiforova, A.M. Petrov, M.M. Basova, and others [11–15]. The green economy is the economy of the future, confirming the importance and relevance of the research topic.

5 Conclusion

Analyzing all of the above, it can be concluded that the development of the "green" economy in small businesses is just beginning. There are numerous obstacles that make this development challenging, such as low environmental responsibility, insufficient government support, costs associated with reducing emissions and waste, the transition away from non-eco-friendly materials, and more. However, it is worth noting that the first steps towards environmentally friendly business practices have already been taken.

The successful transition depends on many factors such as investment, development of partnerships with other businesses, adherence to high product quality standards, and environmental responsibility. Only by adhering to all of these factors can small businesses create long-term value for themselves and the environment.

The role of small businesses is significant as they are drivers of innovation and flexibility and have the potential to benefit the environment and create sustainable business models. By transitioning to the "green" economy, small businesses can minimize emissions of pollutants, control and manage waste, use eco-friendly materials in production, and create environmentally friendly products or services. These steps contribute to the preservation of biodiversity, reduce the negative impact of businesses on the environment, and enhance their image and competitiveness, which undoubtedly affects the economic aspect of the enterprise.

Transitioning to a "green" course creates new business models that have not only positive environmental factors but also economic ones, including the creation of new jobs and increased competitiveness. Such business models align with global sustainable development goals and promote the alignment of business, society, and nature.

References

1. Ivanova, N.I., Levchenko, L.V.: "Green" Economy: Essence, Principles, and Prospects. Bulletin of OMSK State Univ. Ser. Econ. **2**, 19–28 (2017)
2. Smagulova, Z.B., Mukhanova, A.E., Musaeva, G.I.: Analysis of the world experience in transition to a green economy: preconditions and directions. Int. J. Appl. Fundam. Res. **1–1**, 92–96 (2015)

3. Ilyakova, I.E.: Introduction of best available technologies as a factor in the formation of the "Green" economy: an institutional perspective. Nat. Secur. J./nota bene. (5), 30–40 (2021). https://doi.org/10.7256/2454-0668.2021.5.34765, https://nbpublish.com/library_read_article.php?id=34765
4. Kozlova, O.A., Volkova, I.A. Trends in the Development of Green Business: Experience of EU Countries and Russian Practices. Moscow: Znanie-M. 130 (2020)
5. Efremov, N.A., Efremova, M.P.: Global Experience in the Application of "Green" Technologies. Oeconomia et Jus, 1 (2021). https://cyberleninka.ru/article/n/mirovoy-opyt-primeneniya-zelenyh-tehnologiy. Accessed 14 Sep 2023
6. Efremova, M.P., Sakmarova, L.A.: Environmental Entrepreneurship Abroad. Actual Issues of Economics: Collection of Scientific Papers. Cheboksary: Chuvash University Publishing House, pp. 71–74 (2020)
7. Kodaneva, S.I.: Key directions and prospects of energy transition in Russia. Econ. Soc. Issues Russia **1**, 79–94 (2022)
8. Sycheva, A.A.: Features of Implementing Environmental Business Innovations in Foreign Countries. Stolypin Herald, 5 (2022). https://cyberleninka.ru/article/n/osobennosti-vnedreniya-ekologicheskih-biznes-innovatsiy-v-zarubezhnyh-stranah. Accessed 14 Sep 2023
9. Krysanova, N.V.: Review of the book: Kodaneva, S. I. Green Economy - From Understanding the Concept to Practicing Its Implementation (Experience of Russia and Foreign Countries). Social and Humanities Sciences. Domestic and Foreign Literature. Series 4, State and Law: Abstract Journal, **3**, 165–171 (2021)
10. Stepanova, T., Schnaider, O.: The influence of social indicators of economic actors on the construction of an effective ecosystem. Ecosystems Without Borders: Opportunities and Challenges. Conference proceedings. Springer Nature Switzerland AG, pp. 9–18 (2022). https://doi.org/10.1007/978-3-031-05778-6
11. Lipinsky, D.A., Berdnikova, L.F., Schnaider, O.V.: Specific features of training of law makers with the help of remote technologies. In: Popkova, E., Sergi, B. (eds.) Digital Economy: Complexity and Variety vs. Rationality. ISC 2019. LNNS, vol. 87, pp. 606–611 Springer, Cham (2020). https://doi.org/10.1007/978-3-030-29586-8_70
12. Efimova O.V., Nikiforova E.V., Basova M.M., Shnaider O.V., Ushanov I.G.: Practice of non-financial reporting disclosure by Russian companies: bridging the gap between company disclosures on sustainability and stakeholders' needs. In: The Collection: ACM International Conference Proceeding Series. Proceedings of the 5th International Conference on Engineering and MIS, ICEMIS (2019)
13. Berdnikova, L.F., Sherstobitova, A.A., Schnaider, O.V., Mikhalenok, N.O., Medvedeva, O.E.: Smart university: assessment models for resources and economic potential. Collect. Smart Innovation, Syst. Technol. **144**, 583–593 (2019)
14. Petrov A.M., Churilova E.Y., Nikiforova E.V.: Cluster analysis of Russian regions by economic and demographic indicators to improve the strategies of development institutions. Braz. J. Law Int. Relat. **3**(36), 623–647 (2022)
15. Shnaider, V.V., Pipko, E.G., Petrov, A.M.: Current information needs of passenger and freight transportation enterprises in Russia. Bull. Nat. Acad. Sci. Repub. Kaz. **4**(386), 348–353 (2020)

Customer Choice in the Face of Uncertainty

Maria Konovalova[1], Lenar Safiullin[2], and Tatyana Stepanova[3](✉)

[1] Department of Economic Theory, Samara State University of Economics, Samara, Russia
[2] Institute of Management, Economics and Finance, Kazan (Volga Region), Federal University, Kazan, Russia
[3] Department of Economic Security, Kaliningrad State Technical University, Kaliningrad, Russia
tatyana.stepanova@klgtu.ru

Abstract. The article examines the issues of the effectiveness of the functioning and development of economic systems at various levels (national, regional, municipal, sectoral, etc.) and notes their dependence on the intensity and sustainability of consumer choice. Generally, solvent consumer demand creates conditions for generating revenue and profits of business entities of various profiles and, as a result, for maximizing their market value, implementing corporate social programs, full and timely deduction of taxes and other mandatory payments to budgets of various levels. The aim of the study is to develop the theoretical and methodological provisions of the theory of consumer choice in conditions of uncertainty. The work clarifies the mechanism of consumer choice, examines the structure of consumer choice of individuals in the global economy. It is concluded that it is advisable to use a synthesis of theoretical and methodological provisions and methodological tools of certain areas of neoclassical economic theory, economics of neo-institutionalism, as well as a number of provisions of post-neoclassical synthesis, primarily marketing theory in the study of trends and problems of consumer choice development.

Keywords: Theory of Public Welfare · Consumer Surplus · Producer Surplus · Uncertainty · Transaction Costs · Transformational Costs

1 Introduction

The nature of consumer choice in various markets of goods, works, and services directly depends on the degree of relative certainty or uncertainty of financial and economic development. The existence of a close relationship between the level of uncertainty and the usefulness of consumer choice was pointed out by G. Simon. The theoretical view of the influence of uncertainty on the change in the general and marginal utility of consumer choice was argued, in particular, by such representatives of post-neoclassical economic theory as J. Ken and A. Michael. However, the earliest works in the field of consumer behavior are associated with the works of representatives of the Austrian school K. Merger, as well as the English scientist I. Bentham, who first introduced the category of utility into scientific circulation. It should be noted that there are quite

R. Polyakov (Ed.): EcoSyst 2023, LNNS 1092, pp. 65–74, 2024.
https://doi.org/10.1007/978-3-031-67354-2_8

diverse interpretations of the definition of utility, which has undergone a fairly long evolution over several centuries, starting from its interpretation by A.R. J. Turgot, E. B. Condillac and ending with representatives of neo-institutionalism. The noted authors attempted to link the categories of utility and value into a single system, focusing on the predominant role of the utility of the good. As for the contribution of the Austrian school of scientific thought to the development of the theory of consumer behavior, it is difficult to overestimate it. Scientists of this field can be called apologists of the scientific doctrine of the behavior of individuals in the consumer goods market from the standpoint of the subjective psychological system of value and value of goods [1].

The undoubted merit of the Austrians is the development of key principles of utility analysis, not only qualitative, but also quantitative. K. Merger convincingly proved the difference in the usefulness of the same benefits for different categories of individuals, which depends on the preferences of each of them. It is impossible to ignore the Marxist concept, which, however, had its own vision regarding the category of utility. The central definition of the Marxist paradigm of historical development was the category of value as an objective ability to satisfy human needs, measured by labor costs, embodied in a specific product.

2 Methodology

The conducted research is based on the use of basic methodological approaches, as well as specific local methods, among which analysis, synthesis, induction, deduction, and a graphical method of material expression can be noted. A special place among the tools and methods used belongs to the causal method, the use of which allows you to uncover the most profound cause-and-effect relationships between various phenomena and events.

3 Results

As shown in Fig. 1, both the overall and marginal utility of consumer choice decreases, under conditions of uncertainty of the external financial and economic environment, as a rule. The relative reduction in the level of marginal utility of consumer choice in conditions of uncertainty is evidenced by the fact that the function "b" shown in Fig. 1 is relatively less convex compared to the function of type "a", characterized by a less significant marginal effect of the intensity of consumer choice on the likely increase in the overall utility of consumption of certain types of economic benefits.

In particular, within the framework of the residential real estate market, the decrease in the level of usefulness of consumer choice in a situation of prolonged uncertainty caused, for example, by the impact of the macroeconomic crisis can be explained by a set of the following main reasons:

a) reduction in the activity of purchases of residential real estate at an acceptable cost due to the uncertainty of potential consumers in the possibility of further maintenance of the property, timely and full repayment of current mortgage payments, etc., since

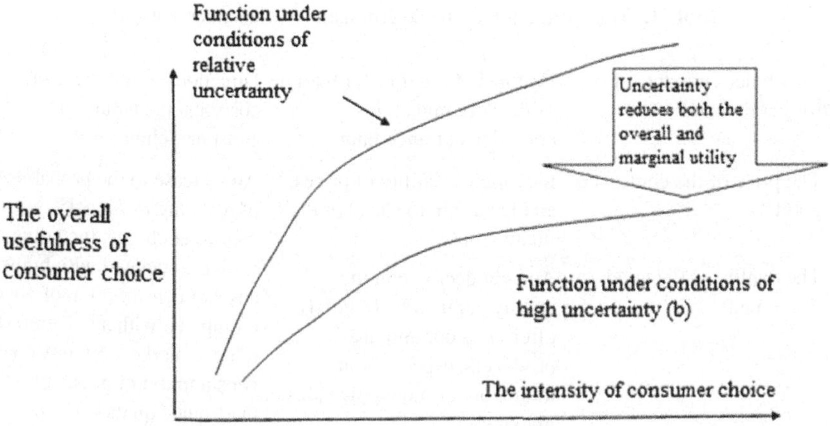

Fig. 1. The impact of uncertainty on the overall and marginal utility of consumer choice (the approach of J. Ken and A. Michael)

in conditions of prolonged uncertainty, the real monetary incomes of most groups of the population are almost always reduced;

b) in conditions of prolonged uncertainty, as a rule, the interest rate on mortgage loans increases - this kind of relationship between the level of uncertainty and the housing mortgage lending market is indicated, in particular, by A.D. Abramov;
c) in a situation of prolonged market uncertainty, the cost of services of effective market intermediaries of the residential real estate market (professional realtors, industry consultants, etc.) usually increases comparatively.;
d) in conditions of prolonged uncertainty and at the same time the comparative weakness of state control institutions, the likelihood of the formation and spread of fraudulent schemes in the residential real estate market increases, fraught for individual subjects of consumer choice with a complete loss of savings generated for the purpose of improving housing conditions [2].

Generally, in conditions of high level of market uncertainty, the nature of the influence of most of the main determinants of demand on the nature and costs of consumer choice is being transformed. The main directions of this kind of transformation are systematized by the author in Table 1.

Directions of influence of the main determinants of consumer demand on costs and the nature of consumer choice in conditions of uncertainty.

The main determinants of consumer demand, the peculiarities of the transformation of determinants in conditions of uncertainty, the directions of influence on costs and the nature of consumer choice.

1. The price of the consumed good. An increase in price volatility, an increase in the level of their fluctuations. An increase in the probability of realizing irrational consumer choice, including a choice associated with higher costs of consumption of goods (compared with a situation of relative market certainty) or consumption of goods of inadequate quality.

Table 1. Transformation of the determinants of consumer demand

The main determinants of consumer demand	Features of the transformation of the determinant in conditions of uncertainty	Directions of influence on costs and the nature of consumer choice
1. The price of the consumed good	Increased volatility of prices, an increase in the level of their fluctuations	An increase in the probability of realizing irrational consumer choice, including a choice associated with higher costs of consumption of goods (compared with a situation of relative market certainty) or consumption of goods of inadequate quality
2. The quality of the good consumed	A slight decrease in the quality parameters of goods offered by commodity producers, especially in monopoly or monopoly sales markets	
3. Information about the potential object of consumer choice and its possible substitutes	Reducing the relevance of information about objects of potential consumer choice	The growth of the consumer's transaction costs associated with collecting information about the goods consumed and checking the level of its representativeness
4. The economic policy of the state in the industry market and in the field of consumer protection	There is a certain lag in the content of state policy, including in the field of consumer protection, from the rapidly changing market situation	Reducing the level of actual protection of the rights and legitimate interests of consumers of goods, works, and services. Increasing potential legal costs for consumers arising from the need to protect their violated interests
5. Opportunities to raise borrowed funds to finance purchases	In conditions of prolonged uncertainty, the cost of consumer, including mortgage, lending usually increases	The overall costs of consumer choice associated with leveraging are increasing
6. Marketing technologies of producers (or sellers) of economic goods	Increasing the risks of incorrect, unprofitable marketing manipulation for consumers	A decrease in the level of rationality of consumer choice under the influence of marketing methods that are not always ethically correct

2. The quality of the good consumed. A slight decrease in the quality parameters of goods offered by commodity producers, especially in monopolistic or oligopoly sales markets.
3. Information about the potential object of consumer choice and its possible substitutes. Reducing the relevance of information about objects of potential consumer choice.

The growth of the consumer's transaction costs associated with collecting information about the goods consumed and checking the level of its representativeness.

4. The economic policy of the state in the industrial market and in the field of consumer protection. There is a certain lag in the content of state policy, including in the field of consumer protection, from the rapidly changing market situation. Reducing the level of actual protection of the rights and legitimate interests of consumers of goods, works, and services. Increasing potential legal costs for consumers arising from the need to protect their violated interests.
5. The possibility of mobilizing borrowed funds to finance purchases in conditions of prolonged uncertainty, the cost of consumer, including mortgage, lending usually increases. The overall costs of consumer choice associated with leveraged fund mobilization are increasing.
6. Marketing technologies of producers (or sellers) of economic goods. Increasing the risks of incorrect, unprofitable marketing manipulation for consumers. A decrease in the level of rationality of consumer choice under the influence of marketing methods that are not always ethically correct.

In the context of the problem under consideration, I would like to focus on the main directions of the influence of marketing activity on the nature of consumer choice in conditions of relative market uncertainty.

The effectiveness of consumer choice in conditions of uncertainty directly depends on the validity of the directions for improving marketing management [3]. At the same time, the modernization of marketing management systems for companies of various profiles in modern business conditions, one of the main attributes of which is the significant dynamism of industry sales markets, should, in our opinion, be based on the systematic use of the following basic managerial and marketing concepts, mainly formed in the paradigm of post-neoclassical economic theory:

1. The concept of customer-oriented marketing (CRM marketing), according to which the marketing mix should be maximally individualized for narrow segments of the industry market for goods, works, and services. Within the framework of the virtual information space, it is even possible to provide an individual marketing package for each potential or real consumer – an individual.
2. Technologies for using big data (large amounts of data, in this case of a marketing and related nature) to improve the efficiency of the organization and implementation of the company's marketing activities. So, the CRM marketing system of an organization is associated with the processes of constant collection and processing of information about customers, industry markets, marketing intermediaries, etc. A certain part of this kind of marketing information is not entirely relevant, and therefore the problem of verification, verification of the reliability of the information flow is relevant. Subsequently, within the framework of modern big data technologies, relatively representative arrays of marketing information are processed, mainly based on the use of a set of methods of economic and statistical analysis and expert systems.
3. The concept of Internet marketing, which includes the following interrelated tools:
   - a marketing research system for consumer choice of target groups of customers in virtual segments of industry markets for goods, works, and services of the organization;

- a system for generating demand and stimulating demand, including a complex of online advertising of the organization, activities in the field of SMM marketing (marketing activities in the space of modern Internet social networks) and SEO management (the company's activity to promote its own brand or individual product positions within search engines);
- a complex of direct sales based on the use of the potential of the organization's website and, in some cases, special mobile applications;
- sales using the services of virtual marketing intermediaries, primarily marketplaces;
- a set of methodological approaches and analytical tools that provide an objective assessment of the effectiveness of promoting the organization's product positions within the virtual segment of the modern digital economy [4].

4. General theory and practice of the formation and development of flexible, primarily design systems and organizational management systems, including management activities in the field of marketing development of the company. To a large extent, this kind of theory is formed in modern economic conditions based on the use of the potential of the concept of the so-called "turquoise organizations" by Frederick Laloux, which is quite popular and is being implemented relatively intensively in various sectors of the economy.

4 Discussion

In the context of the problem under consideration, I would like to focus on the main directions of the influence of marketing activity on the nature of consumer choice in conditions of relative market uncertainty [5].

The effectiveness of consumer choice in conditions of uncertainty directly depends on the validity of the directions for improving marketing management. At the same time, the modernization of management systems for marketing activities of companies of various profiles in modern business conditions, one of the main attributes of which is the essential dynamism of industry sales markets, should, in our opinion, be based on the systematic use of the following basic managerial and marketing concepts, mainly formed in the pa-the paradigm of post-neoclassical economic theory:

1. The concept of customer-oriented marketing (CRM marketing), according to which the marketing complex should be individualized to the maximum extent for narrow segments of the industry market for goods, works, and services. Within the framework of the virtual information space, it is even possible to provide an individual marketing package for each potential or real consumer – an individual [6].
2. Technologies for using big data (large amounts of data, in this case of a marketing and related nature) to improve the efficiency of the organization and implementation of the company's marketing activities. Thus, the organization's CRM marketing system is associated with the processes of constant collection and processing of information about customers, industry markets, marketing intermediaries, etc. A certain part of this kind of marketing information is not entirely relevant, and therefore the problem of verification, verification of the reliability of the information flow is relevant. Subsequently, within the framework of modern big data technologies, relatively representative arrays of marketing information are processed, mainly based on the use of a set of methods of economic and statistical analysis and expert systems [7].

3. The concept of Internet marketing, which includes the following interrelated tools:
   - a marketing research system for consumer choice of target groups of customers in virtual segments of industry markets for goods, works, and services of the organization;
   - the system of demand formation and demand stimulation (FOSSTIS in the terminology of F. Kotler), which includes a complex of online advertising organizations, activities in the field of SMM marketing (marketing activities in the space of modern Internet social networks) and SEO management (the company's activity to promote its own brand or individual product positions within search engines);
   - a complex of direct sales based on the use of the potential of the organization's website and, in some cases, special mobile applications;
   - sales using the services of virtual marketing intermediaries, primarily marketplaces;
   - a set of methodological approaches and analytical tools that provide an objective assessment of the effectiveness of promoting the organization's product positions within the virtual segment of the modern digital economy.
4. General theory and practice of the formation and development of flexible, primarily design systems and organizational management systems, including management activities in the field of marketing development of the company. To a significant extent, this kind of theory is being formed in modern economic conditions based on the use of the potential of the concept of the so-called "turquoise organizations" by Frederic Laloux, which is quite popular and is being implemented relatively intensively in various sectors of the economy.

The main tools of marketing manipulation of consumer choice processes, which are usually more actively used in conditions of uncertainty, are:

1. Individual customer loyalty programs. In principle, the individual tools of consumer loyalty of companies represent a well-known method of activating consumer demand, systematically worked out in the theory and practice of marketing process management. These categories include individual discounts for different groups of customers, general discounts valid for a certain time period, the issue and subsequent use of discount cards, etc.
2. Coalition consumer loyalty programs. Unlike individual marketing loyalty programs, this rather potentially effective sales intensification tool is relatively widespread in modern industry markets of the national economy of the Russian Federation [8].

It should be noted that in 2015–2022, coalition consumer loyalty programs were fragmentally distributed in the segment of large Russian companies, for example, coalition programs with the participation of Sberbank PJSC and Aeroflot JSC, between separate large banking structures of the national financial and credit system of Russia and federal retail chains, etc. were relatively effectively implemented. At the same time, such programs are practically not implemented in the field of domestic small businesses, as well as in the segment of production and sale of goods belonging to the category of means of production (B2B).

The essence of the coalition consumer loyalty program, which directly affects the nature and costs of consumer choice, is that a consumer who regularly buys goods or services of company 1 in significant amounts receives preferences in terms of purchasing goods or services offered by companies 2 and 3, or participates in marketing campaigns common to the coalition program, such as The rule is reduced to the drawing of valuable prizes or substantial and long-term discounts for the winners of the corresponding drawings [9].

It should be noted that, according to a number of reputable researchers of modern trends and problems of organizing and implementing marketing activities, coalition consumer loyalty programs are a more effective tool for activating demand compared to traditional, individual programs.

Coalition consumer loyalty programs can be operated both directly by their participants and by some independent operator, which can be a fairly reputable marketing or consulting structure. In the latter case, the probability of various kinds of abuses in the process of forming and implementing such a consumer loyalty program decreases, the level of its validity and, accordingly, its potential financial and economic efficiency increases [10].

At the same time, the main functions of the operator of the coalition consumer loyalty program are:

- formation of a general strategy for managing consumer loyalty based on the coalition program;
- development of tactics, immediate conditions and mechanisms for the implementation of the coalition loyalty program, justification of their financial provision;
- financial management of the coalition loyalty program, which are accumulated from the funds of its participants, usually on the basis of parity financial security;
- promotion of the loyalty program in the market of marketing services, aimed at increasing the interest in it of both real and potential target groups of consumers;
- organization of events within the framework of the program (meetings of loyal visitors, issuance of special cards within the framework of the program, carrying out prize draws, etc.);
- monitoring and evaluation of the financial and economic effectiveness of the implementation of the coalition consumer loyalty program.

3. Activation of the use of the potential of marketing intermediaries, including within the virtual segment of the modern digital economy.

It should be noted that the specialized literature presents diametrically opposed approaches to the question of the expediency of using the potential of market intermediaries in the process of influencing the nature of consumer choice in conditions of uncertainty. For example, A.S. Parfenov believes that the services of such intermediaries should be resorted to only in extreme, exceptional cases, since their use not only reduces the level of potential operating profit of the organization, but also is fraught with dangerous managerial consequences, such as, in particular, the leakage of confidential, commercially significant information to which the relevant intermediaries have access. According to the approach of this researcher, accordingly, any form of direct marketing is strategically and tactically more effective than using the capabilities of the institute of market intermediaries [11].

At the same time, a different approach is dominant in the theory of influence on consumer choice processes (according to which the use of the potential of market intermediaries of various types is acceptable even to the extent that the marginal efficiency from the use of their services is positive. At the same time, of course, before interacting with marketing contractors, it is advisable to conduct a comprehensive study of the previous activity of the latter in terms of evaluating professionalism and goodwill [12].

Generally, the range of possible market intermediaries used by the company, subject to the criteria of potential effectiveness, professionalism and integrity highlighted above in terms of fulfilling contractual obligations, including on issues of non-disclosure of confidential information about the company's activities, should be quite extensive, which will allow the institute to make the most diverse use of the potential of the institute mediation. At the same time, the main types of market intermediaries in modern business conditions are:

- dealers of the company that ensure the promotion of products or services in local segments of the industry market;
- franchisors (a typical partner for some large companies with fairly well-known brands) who carry out commercial activities for brand development in their own interests and at their own risk, based on mechanisms for the acquisition of franchises and subsequent support;
- traditional-type commodity exchanges;
- marketplaces on the Internet (Ozone, Amazon, Yandex-market, Ali-express, etc.) that promote mainly consumer goods through their own digital platforms, which, as a rule, have a wide level of fame within the virtual information space;
- remote transport and logistics centers.

4. Activities in the field of SEO management, which is the purposeful activity of the company's marketers to promote it in general or individual product positions in the space of search engines.
5. SMM marketing, which is a system of measures to ensure and increase consumer loyalty in the information space of social networks (VKontakte, Instagram, Tiktok, Facebook, etc.)

5 Conclusion

In conditions of a high level of uncertainty, the influence of marketing tools on the nature, including costs and usefulness, of consumer choice is not completely unambiguous. On the one hand, rational loyalty programs reduce the potential costs of consumers; on the other hand, in conditions of uncertainty, the likelihood of the formation and implementation of incorrect marketing manipulation schemes increases, which negatively affect the effectiveness of consumer choice in general.

References

1. Krestyanpol, L.: Simulation modeling of consumer behavior within the concept of smart consumption. Procedia Comput. Sci. **217**, 774–783 (2023). https://doi.org/10.1016/j.procs.2022.12.274

2. Pandey, S.: Joint inventory and transshipment decisions with consumer behavioral. Appl. Soft Comput. **142**, 110356 (2023). https://doi.org/10.1016/j.ijar.2023.108938
3. Integrated preference argumentation and applications in consumerbehaviour: Hung Nguyen Duy. Procedia Comput. Sci. **221**, 797–804 (2023). https://doi.org/10.1016/j.procs.2023.08.053
4. Weinrich, R.: Leveraging strategic consumer behavior in consumer-to-consumer resale: From economic and environmental perspectives **89**, 102732 (2023). https://doi.org/10.1016/j.envdev.2023.100890
5. Vidal-Ayuso, F.: The circular economy and consumer behaviour: literature review and research directions. Econ. Anal. Policy **71**, 180–197 (2021). https://doi.org/10.1016/j.jclepro.2023.137824
6. Vranken, L.: Simulation modeling of consumer behavior within the concept of smart consumption. North Am. J. Econ. Finan. **58**, 101525 (2023). https://doi.org/10.1016/j.procs.2022.12.274
7. Saidani, M.: Consumer behavior in the circular economy: developing a product-centric framework. Energy Econ. **126**, 106891 (2023). https://doi.org/10.1016/j.jclepro.2022.135568
8. Khatami, F.: Res. Int. Bus. Financ. **66**, 101989 (2023). https://doi.org/10.1016/j.jenvman.2022.116555
9. Alyahya, M.: Understanding the factors affecting consumers' behaviour when purchasing refurbished products:. Resources Policy **72**, 102083 (2021). https://doi.org/10.1016/j.jretconser.2023.103492
10. Evan Weingarten.: Nostalgia and consumer behaviorh. Resour. Policy **70**, 101887 (2021). https://doi.org/10.1016/j.copsyc.2022.101555
11. Li, X.: Exploring the factors influencing consumer engagement behavior regarding short-form video advertising: a big data perspective. Finan. Res. Lett. **57**, 104202 (2023) https://doi.org/10.1016/j.jretconser.2022.103170
12. Zhang, Y.: Behavior-Based pricing under asymmetric channel competition with strategic consumers. Am. J. Econ. Finan. **58**, 101497 (2021). https://doi.org/10.1016/j.omega.2023.102916
13. Siegrist, M.: The application of virtual reality in food consumer behavior research: a systematic review. Energy Rep. **7**, 693–703 (2021). https://doi.org/10.1016/j.tifs.2021.07.015
14. Systematized by the author on the basis of information provided in: Becker G. A Theory of the Allocation of Time, – Economic Journal, 1995, Vol. 75, No. 299. Sept. P. 493; Hodgson, J. Economic Theory and Institutions: A Manifesto of Modern Institutional Economic Theory / J. Hodgson // Translated from English – M.: Delo, 2020; Angel, D.F. Consumer behavior / D.F. Angel, R.D. Blackwell, P.W. Miniard. – St. Petersburg: Peter kom (1999). https://doi.org/10.1016/j.copsyc.2022.101555

Circular Economy and Recycling as Effective Tools of Green Economy

Tatiana Izzuka[1](✉), Natalya Smagina[2], Tatiana Petrusevich[1], and Svetlana Milovidova[1]

[1] FinancialUniversity under the Government of the Russian Federation, Moscow, Russia
tbizzuka@fa.ru
[2] MGIMO University, Moscow, Russia

Abstract. The rapid development of modern society and the scientific and technological progress that has been taking place worldwide since the beginning of the 20th century have led to an increase in the negative anthropogenic impact on the environment. Thus, one of the most significant challenges facing humanity is global warming. The increase in the average global temperature is a critical issue, causing a range of global effects such as the melting of glaciers, disruption of the Earth's thermos circulation model, and rising sea levels. The danger of these effects lies in their consequences, including wildfires, droughts, floods, threats to biodiversity, and so on. Global warming is a result of the intensification of the greenhouse effect, caused by the emissions of greenhouse gases into the atmosphere, such as carbon dioxide (CO2), nitrogen oxides, water vapor, fluorine compounds, chlorine, and so on, many of which are produced by industrial enterprises. Some of these emissions are released directly into the atmosphere, while others are generated during the decomposition of production waste. Mitigating the negative anthropogenic impact of industrial activities can be achieved through the implementation of green economy principles, including circular economy. This article explores the tool of green economy, specifically circular economy and its component - recycling. It identifies the key aspects, benefits, and the necessity of adopting a closed-loop economy for sustainable development. The international experience is examined, as well as the prerequisites for the widespread adoption of this practice in our country.

Keywords: Sustainable Development · Global Warming · Greenhouse Effect · Green Economy · Circular Economy · Closed-loop Economy · Recycling

1 Introduction

In the current realities of human development, global climate change takes center stage among the various problems and challenges facing all countries worldwide. The cause of these changes is the exponential growth of scientific and technological progress in the 20th and 21st centuries, and as a result, the increasing anthropogenic and technogenic negative impact on the environment. Due to this impact, there is an observed increase in the average global temperature. For example, in 2022, the global average temperature

R. Polyakov (Ed.): EcoSyst 2023, LNNS 1092, pp. 75–83, 2024.
https://doi.org/10.1007/978-3-031-67354-2_9

exceeded the pre-industrial period (1850–1900) level by 1.15 ± 0.13 °C [1]. According to scientists' forecasts, by the year 2100, this indicator is expected to exceed the pre-industrial level by 4.5 °C, indicating an unprecedented rate of increase. Such a global average temperature was last seen on Earth approximately 20 million years ago [2], when the first ancient hominoid apes appeared, marking the beginning of our species' history. The specific impact of technological progress over the last 30 years on the annual average temperature can be visualized through a temperature change map (see Fig. 1).

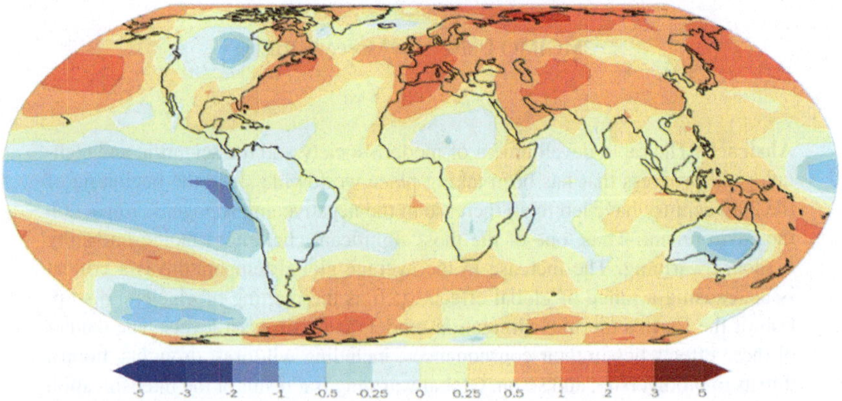

Fig. 1. The difference in annual average temperature between 1991 and 2022 [1].

The global trend of global warming, according to scientists, is primarily attributed to the intensifying greenhouse effect, which disrupts the planet's temperature regulation [3]. This effect is explained by the presence of greenhouse gases in the Earth's atmosphere, which hinder the escape of heat back into space, heat that the Earth receives from the Sun. The increasing concentration of these gases in the atmosphere leads to a rise in the global average temperature. Greenhouse gases include carbon dioxide, methane, nitrogen oxides, fluorine and chlorine compounds, as well as water vapor. The total emissions of these gases (see Fig. 2) in recent years have already exceeded 50 gigatons in terms of carbon dioxide equivalent (GtCO2e).

The intensifying greenhouse effect and, consequently, global warming have compelled all countries worldwide to reevaluate the trajectories of their development and recognize the necessity for sustainability. In 1992, at the United Nations Conference held in Rio de Janeiro, the "Agenda for the 21st Century" was formulated, outlining the goals of sustainable development. In 2015, 193 United Nations member states committed to achieving these goals [5, 6]. One of the sustainable development goals is Goal 12, "Responsible Production and Consumption", aimed at reducing greenhouse gas emissions from production by implementing the principles of the green economy.

One of the effective directions of the green economy is the circular economy or closed-loop economy, based on the reintegration of waste, by-products, and resources into the production cycle of enterprises. In this context, recycling should be considered as one of the promising ways to achieve production sustainability and, consequently, reduce the negative impact on the environment.

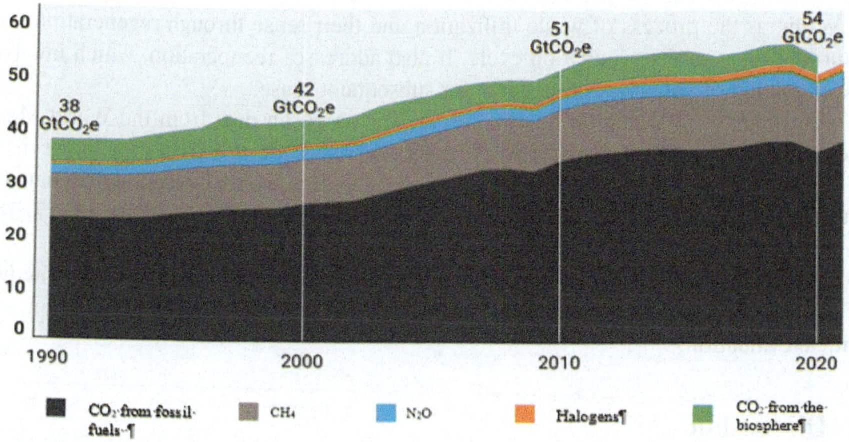

Fig. 2. Change in greenhouse gas concentrations from 1990 to 2021, gigatons in CO_2-equivalent ($GtCO_2e$) [4].

2 Methods and Materials

Green economy is a resource-efficient economy based on low-carbon development, which leads to the improvement of people's well-being and social justice, while reducing environmental risks and preventing the loss of biodiversity [7].

Green economy is a direction in economic science that has emerged in the last two decades, within which it is considered that the economy is an interdependent component of the natural environment, within the boundaries of which it exists and is an integral part of it [8].

Additionally, let's explore one of the directions within the green economy, which is circular economy.

Circular economy is a new direction in economics aimed at addressing the environmental crisis and the depletion of natural resources by transitioning to a more efficient resource utilization system. The implementation of a closed-loop economy requires a shift in thinking and production approaches, as well as the development of new technologies and infrastructure [9].

The technological support of its principles plays a crucial role in the implementation of the circular economy concept, with recycling technologies taking a prominent position [10].

One of the fundamental components of the green and circular economy is recycling.

In foreign practice, the definition of recycling is characterized as the process of secondary processing of waste generated during human society's activities. The primary goal of recycling is to reduce waste disposal volumes, aiming to prevent the loss of potentially valuable materials. This, in turn, leads to a reduction in the consumption of primary resources and energy, as well as a less aggressive impact on nature and the environment.

In Russian practice, the legislative clarification of waste management is provided by Federal Law No.89-FL "On Production and Consumption Waste" [11].This law defines recycling as the process of waste utilization and their reuse through regeneration, i.e., returning them to the production cycle. It also addresses recuperation, which involves the extraction of valuable components for subsequent reuse.

The scientific foundation for this research is based on data from the World Meteorological Organization, the United Nations, and the Federal Service for Supervision of Natural Resource Usage (Rosprirodnadzor). Scientific works by leading Russian and international experts in the field of green and circular economy [2, 9, 10, 12, 17, 18], among others, were used as references.

The objectives of this research were achieved through the content analysis method of scientific publications and reports, followed by data examination, organization, and critical evaluation.

3 Discussion

One of the factors of sustainable development is the green economy, which contributes to minimizing the negative impact on the environment and promotes the efficient use of both natural and non-natural resources. "The scope of tasks in the green economy is defined by inclusive growth in terms of rational natural resource utilization, reducing sources of hydrocarbon, minimizing emissions into the atmosphere, and preserving ecosystems" [12]. Both the green economy and the circular economy are key components of sustainable development, aimed at creating efficient systems of production and consumption, pursuing a common goal, and contributing to the improvement of the ecological situation. Let's take a closer look at the main aspects of these economic directions (see Fig. 3).

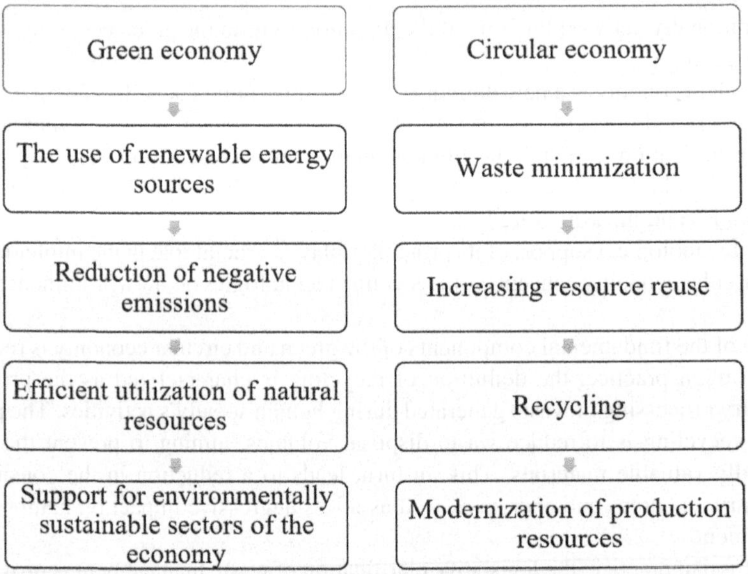

Fig. 3. Aspects of the green and circular economy.

Based on this Fig. 3, one can draw conclusions about the interrelation of these economies. (see Fig. 4).

Resource optimization

- The circular economy addresses one of the aspects of the green economy in terms of the efficient utilization of natural resources

Minimization of waste

- One of the aspects of the circular economy is recycling and waste processing, which aligns with the principles of the green economy

Sustainable consumption

- The circular economy promotes ideas of sustainable consumption models, which also align with the goals of the green economy

Fig. 4. The interrelation between the green and circular economies

It is worth noting that the circular economy is an important tool within the framework of the green economy, contributing to the minimization of environmental damage and adhering to the principles of sustainable development.

One of the key principles of the circular economy is the return of production waste, by-products, and resources into the technological chain of enterprises. In this context, recycling should be considered as a primary component of effective waste utilization within a company's production cycle.

As mentioned earlier, recycling is the process of converting waste into reusable materials, which includes the following steps:

- Collection. All production waste is collected in specialized containers, separate from household waste.
- Sorting. Waste is sorted by material type (plastic, paper, glass, etc.).
- Processing. Each sorted waste type is processed accordingly.
- Manufacturing new products or resources. The resulting material is reused for the production of energy, new materials, goods, and more.

The importance of recycling lies in the following:

1 It allows for the reuse of resources, contributing to the conservation of natural resources.
2 Through waste processing, secondary materials are reintroduced into the production cycle, thereby minimizing the negative impact on the environment.
3 It reduces the production of new materials by utilizing old ones, leading to a reduction in environmental pollution.
4 It promotes the creation of new job opportunities in recycling and waste management sectors.

Despite the popularization of the green economy, waste disposal remains one of the global environmental challenges. According to the 2022 report from Rosprirodnadzor, the amount of waste generated in 2021 amounted to 8 448, 6 million tons. In previous years (2019 and 2020), this figure was 7 750, 8 and 6 955, 7 million tons, respectively (see Fig. 5).

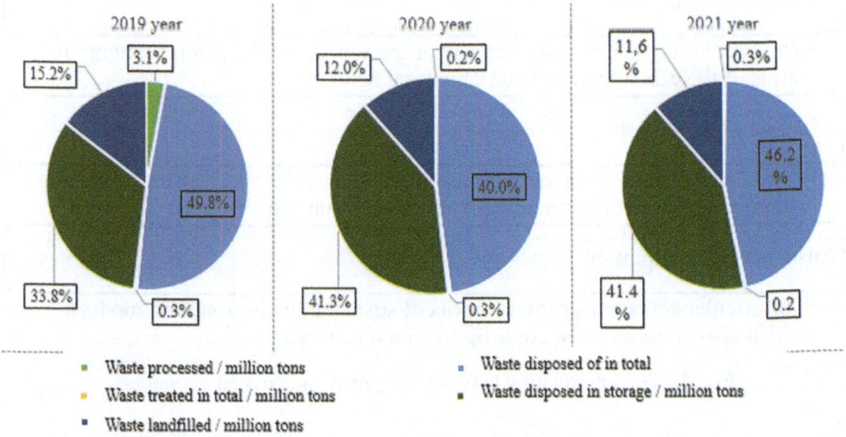

Fig. 5. Information about waste for the years 2019 - 2021 (compiled by the authors based on materials [13]).

Based on the presented data, it is evident that there is an annual increase in the generation of production and consumption waste, which, in turn, has a negative impact on the overall environmental situation. "To address this environmental issue, the state needs to pay attention to the best global practices aimed at combating waste generation" [14] (Table 1).

Table 1. Global practices regarding the disposal of production waste.

Country	Measures for the disposal of production waste
Germany	In the country, a recycling system has been established and is regulated at the state level. Over 60% of the waste is recycled to produce secondary products and generate electricity
The Netherlands	The Ocean Cleanup organization has created a project aimed at cleaning up the world's oceans (collecting and recycling plastic in the ocean). As of now, more than 40 tons of plastic have been extracted from the Pacific Ocean [15]
Canada	The country is moving away from plastic usage, implementing high tariffs on waste exports, and recycling a significant portion of the waste. This allows Canada to hold one of the leading positions among countries in terms of responsible environmental stewardship
France	The reduction of the number of waste incineration plants, with approximately 40% of the waste produced in the country going through recycling

Foreign countries are successfully implementing the principles of the circular economy, effectively addressing environmental issues. Their experience shows that recycling, as one of the main directions of the circular economy, makes a significant contribution to environmental conservation. It not only leads to a reduction in the ecological footprint but also promotes the development of environmental awareness and responsibility among the population.

4 Research Results

The path to sustainable development and the implementation of the circular economy in Russia is quite challenging for a number of reasons:

1 A large number of illegal dumps. In 2022, the country had 16.4 thousand illegal landfills [16].
2 Low level of environmental responsibility among the population. The majority of the country's residents do not fully comprehend the scale of the problem, do not engage in household waste sorting, and do not use products from secondary processing.
3 Insufficient number of specialized professionals in the field of ecology.
4 Most corporations neglect the principles of the circular economy as transitioning to it is a fairly labor-intensive process.

It's essential to address these challenges and promote awareness and responsible practices to foster a circular economy and sustainable development in Russia.

"According to the amount of waste generated and the lack of competence in waste management, the Russian Federation needs an urgent transition to a circular economy, which is achieved through the use of business models" [17].

Let's take a closer look at the five primary business models developed by Accenture (see Fig. 6).

Furthermore, in addition to the application of business models, government support is required when transitioning to a circular economy. In Russia, there are already several policy documents and laws aimed at resolving the environmental situation in the country. These include:

1. Green Standards of the Shanghai Cooperation Organization (SCO).

2. BRICS Green Technology Platform.

3. Federal Law dated July 2, 2021, No. 296-FL "On Limiting Greenhouse Gas Emissions".

4. Article 17 of Federal Law dated January 10, 2002, No. 7-FL "On Environmental Protection".

5. Decision of the Eurasian Economic Commission dated November 23, 2020, No. 116 "On the Development Program of Integration in the Sphere of Statistics of the Eurasian Economic Union for 2021–2025" and others.

However, despite the existing legislative initiatives in this area, there is a lack of a unified and comprehensive government support system for the effective and widespread transition of Russian enterprises to the circular economy model.

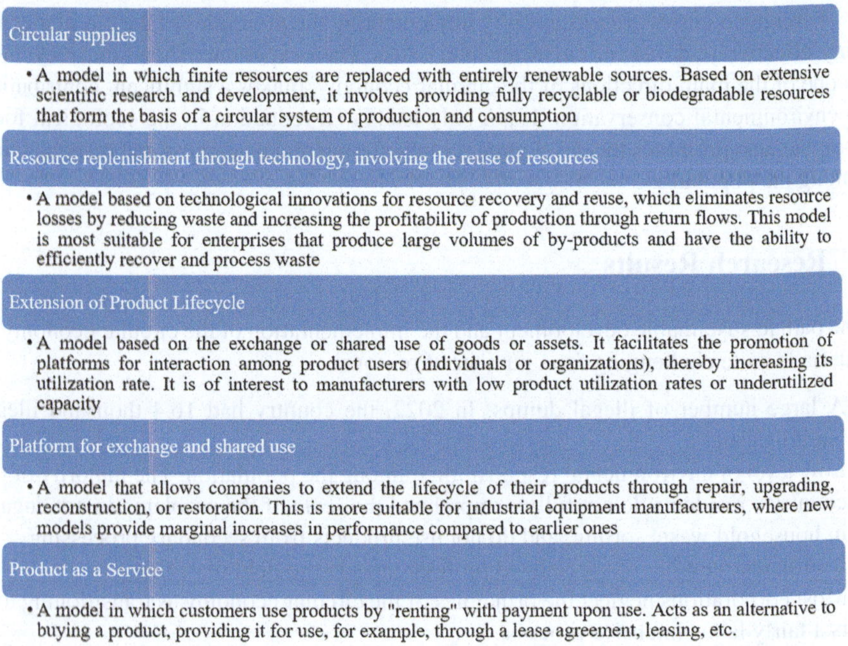

Fig. 6. Five core business models developed by Accenture (compiled by the authors based on materials from [18]).

5 Conclusion

The results of this research emphasize the necessity for Russia to transition to circular economy business models. Existing waste management methods not only fail to improve the environment but also pose a threat to the ecosystem. Despite the numerous challenges, this transition is a complex and long-term process, as the established system in Russia requires radical changes at all levels, from government to the general population. Such a transition would contribute to the creation of a new responsible culture of consumption and production, directly impacting environmental preservation and the well-being of the country's population and the planet.

References

1. World Meteorological Organization: Past eight years confirmed to be the eight warmest on record, Press release of 12 January 2023. https://public.wmo.int/en/media/press-release/past-eight-years-confirmed-be-eight-warmest-record. Accessed 04 Oct 2023
2. Burke, K.D., et al: Pliocene and Eocene provide best analogs for near-future climates. Proc. Nat. Acad. Sci. U.S.A. **115**(52), 13288–13293 (2018)
3. IPCC, 2021: Summary for policymakers. In: Climate Change 2021: The Physical Science Basis. Contribution of Working Group I to the Sixth Assessment Report of the Intergovernmental Panel on Climate Change. In: Masson-Delmotte, V., et al. (eds.). Cambridge University Press, Cambridge, United Kingdom and New York, NY, USA, pp. 3–32. https://www.ipcc.ch/report/ar6/wg1/downloads/report/IPCC_AR6_WGI_SPM.pdf. Accessed 04 Oct 2023

4. The Closing Window-Climate Crisis Calls for Rapid Transformation of Societies: Emissions gap report 2022, United Nations environment programme. https://www.unep.org/resources/emissions-gap-report-2022/ Accessed 04 Oct 2023
5. United Nations: Agenda 21 – "The Agenda for the 21st Century" (1992). https://sustainabledevelopment.un.org/content/documents/Agenda21.pdf. Accessed 04 Oct 2023
6. United Nations: Transforming our world: the 2030 agenda for sustainable development. resolution adopted by the United Nations general assembly on September 25 (2015). https://documents-dds-ny.un.org/doc/UNDOC/GEN/N15/291/92/PDF/N1529192.pdf?OpenElement. Accessed 04 Oct 2023
7. United Nations Environment Programme (UNEP): Established by the United Nations general assembly on December 15 (1972)
8. Chovgan, N.I., Akupiyan, O.S.: Investment support for the agricultural sector of the economy: problems and solutions. Russian Econ. Online J. **4**, 122 (2018)
9. Belyakova, E.A.: The concept of implementing circular economy. Bull. Sci. **5**(62), 1186–1190 (2023)
10. Smirnov, V.V.: Recycling: a strategic direction for enhancing efficiency in metallurgy in the Russian federation. Account. Anal. Audit **5**(4), 30–39 (2018)
11. Federal Law: On production and consumption waste dated June 24, No. 89-FL (as amended on August 4, 2023, No. 476-FL) (1998)
12. Schnaider, V.V.: The role of green economy in achieving sustainable development. Econ. Manag. **2**(43), 71–73 (2023)
13. Report on the activities of the federal service for supervision in the sphere of natural resource use in 2022. https://rpn.gov.ru/upload/iblock/92c/94e0t5bajsoxd1l947ompu2mphykcsw7/Doklad-2022-dop2.pdf. Accessed 04 Oct 2023
14. Kozlova, Y.A., Noskov-Dukelski, A.I.: The Problem of Environmental Pollution in Russia by Production and Consumption Waste, Symbol of Science, 3–2, 8–12. https://cyberleninka.ru/article/n/problema-zagryazneniya-prirodnoy-sredy-rossii-othodami-proizvodstva-i-potrebleniya. Accessed 04 Oct 2023
15. How Clouds Help Solve the Problem of Waste Disposal, Habr. https://habr.com/ru/company/cloud_mts/blog/573946/. Accessed 04 Oct 2023
16. Regions with the Highest Number of Illegal Dumpsites Named, FinExpertiza - A Network of Professional Audit, Appraisal, and Consulting Companies. https://finexpertiza.ru/press-service/researches/2023/reg-naibolsh-chisl-sval/. Accessed 04 Oct 2023
17. Koroleva, E.D., Kolchina, V.V.: Circular economy: domestic experience, key business models, and challenges of their implementation in the Russian federation. Bull. Youth Sci. **2**(29), 1 (2023)
18. Aleksandrova, V.D.: Business models of circular economy. Int. J. Humnit. Nat. Sci. **5-1**, 94–97 (2019)

Spatial Ecosystems

Spatial Transformation of Organic Farming in the Sense of Innovation Theory by Hägerstrand

Denis Galkin(✉)

Altai State Agricultural University, Barnaul, Russia
gennadichu@mail.ru

Abstract. We assume that the spatial diffusion of organic agriculture can be corresponded to the stages of T. Hägerstrand's theory of spatial diffusion of innovations. This allows us to assess the homogeneity of the distribution of organic agriculture and divide all producers into two main groups: innovators and followers. Our previous study of general population of organic products producers in Russia made it possible to identify and evaluate each stage of the spatial diffusion of organic agriculture. At the initial stage, about half of the producers of organic products were certified according to international standards and were located near the Moscow region. In the current paper, we update our previous results from the diffusion stage according to the new 2023 data. We observe the appearance of followers certified according to national standards. Currently, the share of manufacturers certified according to Russian standards reached 84.6%. Followers are located mainly in the central as well as south parts of the country and are concentrated around sales markets. We point out that the development of the organic sector remains spatially heterogeneous. We think that this is due to insufficient effective demand, the lack of targeted government support measures, and flaw of experience among small farmers. Therefore, in the short term (2020–2023), even the intensity of the appearance of followers is not observed. We conclude that scaling up organic agriculture and moving to the densification stage is only possible in the long term and will be determined by the possibilities of government support.

Keywords: Organic Agriculture · Location · Region · Specialization · Demand Factors · Innovators · Followers

1 Introduction

We consider organic agriculture as an innovation and it can be developed in similar stages, for example: (1) initial idea, (2) diffusion and (3) densification [1, 2]. Previously, we compared the stages of organic agriculture in Russia and the spatial diffusion of innovations by T. Hägerstrand. This comparison made it possible to determine the main stages of the placement of organizations in the sector, and to divide producers of organic products into *innovators* and *followers*. The results of the initial stage research showed (1) the concentration of innovators around cities or agglomerations with a high level of

R. Polyakov (Ed.): EcoSyst 2023, LNNS 1092, pp. 87–94, 2024.
https://doi.org/10.1007/978-3-031-67354-2_10

innovation potential and (2) the emergence of followers. At the initial stage, there was a concentration of innovators in the central part of the country and Western Siberia (see Fig. 1) [3]. However, in our interest are the spatial transformation of organic agriculture and the assessment of the factors that influencing it. To do this, we place organizations certified according to organic standards in 2023 on a geographic map and try to assess the changes that have occurred.

Fig. 1. Location of organizations certified according to organic standards in 2020

In the results section of the present article, we determine the effectiveness of organizations in each specialization. We also reveal the key factors in the spatial transformation of organic agriculture include geopolitical and economic ones. As part of the economic factor, we study demanding factors: (1) prices for substitute goods, (2) income, (3) consumer preferences, (4) the number of consumers and (5) consumer expectations. Based on the analysis of factors, we identify the possibilities: (1) increasing retail sales in the domestic market, (2) increasing production volumes, (3) changing the number of producers of organic products.

2 Methods: The Stages of T. Hägerstrand's Theory and Spatial Transformation of Organic Agriculture

T. Hägerstrand produced a theory with initial and diffusion stages of organic innovation [4]. We propose to use mapping research method to assess the location of organic certified organizations. At the initial stage, we analyzed 2020 data to identify innovators and assess their placement [3]. At the diffusion stage, we identified followers and their spatial distribution was assessed based on data from 2023. When plotting innovators and followers on the map, we are able to use a continuous method of statistical observation; some organizations from the lists had national and international certificates. For the statistical basis for the study, we obtained the data from the Ministry of Agriculture of the Russian Federation, as well the data from the Union of Organic Farming [5], and reports from major producers of organic products [6].

To assess the factors influencing the spatial transformation of organic agriculture, a statistical grouping was carried out according to two characteristics: size and specialization. Small and medium-sized organizations specializing in «Cultivation and processing of crops», «Collection of wild fruits and berries, their preservation», «Production of alcoholic beverages», «Mixed specialization (crop and livestock), food production», «Production of meat, milk and dairy food products», «Exporters and processing organizations».

3 Results and Discussions

3.1 Spatial Diffusion and Efficiency of Organic Agriculture

According to T. Hägerstrand's theory [4], the appearance of followers should be observed at the «diffusion» stage. In 2023, 84.6% of organizations in the general population are certified according to Russian standards, the remaining 15.4% according to international ones. Followers, as a rule, appear in territories adjacent to Moscow, the Moscow and Tomsk regions (neighborhood effect), as well as the southern regions of the country (see Fig. 2).

Fig. 2. Placement of organizations certified according to organic standards in 2023

The largest decrease in innovators was observed in Moscow (by 33% to 8) and the Moscow region (by 14.2% to 12). Innovators from the Tomsk region, that are certified according to international standards, continued to operate in 2023. In 2023, the agro-industrial holding LLC «Ekonova-Apk Holding» (an innovator) remained with 2 organizations with organic certificates: «Savinskaya Niva» LLC, «Severnaya Niva Organic» LLC. The largest increase in followers in 2023 relative to 2020 was observed in the

Krasnodar Territory (by 87.5% to 15 organizations), Voronezh Region (2.8 times to 14), Novosibirsk Region (4 times to 8), Primorsky Territory (4 times to 4).

In 2023, several specializations in organic agriculture were noticed. We observed different producers groups and calculated the share of producers based on the specialization. Eventually, 35.9% of the total number of certified organizations is specialized in crop production. Most of the plant growing organizations are located in the Krasnodar Territory and Tomsk Region. A large number of followers were observed in the south and central part of Russia. Crop innovators remained mainly in the Tomsk region, and in 2023 continued to produce products in accordance with international standards. The share of mixed agricultural organizations in 2023 was 9.4% of the total population, most of them are located in the Moscow, Yaroslavl and Novosibirsk regions. The share of alcoholic beverage producers was 8.8% - these are organizations located in the Moscow and Kemerovo regions, Altai Territory, St. Petersburg, and the Republic of Mordovia. 7.1% of organizations that are located in the Yaroslavl region specialized in the production of meat and milk. The share of procurers and processors of wild fruits and berries in 2023 was 3.5%. This category was dominated by organizations from Siberia (Tomsk region, Altai and Krasnoyarsk territories, Republic of Buryatia), as well as from the Leningrad region and Krasnodar region. The number of certified organizations specializing in the collection of wild fruits and berries in 2023 compared to 2020 decreased by more than 2 times. This is due to the export orientation of organizations and a decrease in the volume of supplies of organic products to the European market.

In Table 1, we split the effectiveness of small and medium-sized organizations certified according to organic standards. The type of activity «production of alcoholic beverages» was not taken into account, since it is represented by large organizations with a small share of organic production. We use the 2022 information base of organizations reports. Data from small farms that do not publish reports were not taken into account. When assessing the performance of organizations, we taken into account the values of ROM (Return on Margin), ROS (Return On Sales), ROA (Return on Assets). For the types of activities «Mixed agriculture», «Collection of wild fruits and berries, conservation», «Production of meat and milk», data from only one of medium-sized organization were taken into account.

Organizations with maximum ROM values specialized in mixed agriculture («Biosphere» LLC, «Savinskaya Niva» LLC, «Sib-Kolos» LLC, « Tul'skiy zveroboy» LLC, «Belye Lugi» Peasant Farm, «Ryabinki Ecofarm», «Severnaya Niva Organic» LLC, Breeding plant «Rassvet»), as well as collecting wild fruits and berries in: (1) Tomsk region (LLC «Green»), (2) Altai Territory (LLC «From Wild»), (3) Republic of Buryatia («Ayu Group» LLC, «Taiga Trade» LLC), (4) Novosibirsk region («Green Pantry» LLC), (5) Leningrad region («Lenlesprodresurs» LLC). The highest ROS and ROA values are typical for organizations collecting wild fruits and berries.

Organic agriculture in 2023 remains at the «diffusion» stage. There is a decrease in the number of innovators and an increase in followers. In 2020–2023, there is no significant increase in the total number of producers of organic products. This may be due to the fact that in Russia the same government support tools are available to producers of organic products as for producers of non-organic products. Targeted tools to support the production of organic products have not been identified; there is only regional experience with hectare-by-hectare support for organic farms.

Table 1. Efficiency of activities of organizations certified according to organic standards

Activity type	ROM, %	ROS, %	ROA, %
Growing and processing of agricultural crops:			
– small organizations	-4.92	4.37	1.68
– medium-sized organizations	7.02	0.92	1.07
Mixed farming:			
– small organizations	38.08	2.68	0.94
– medium-sized organizations	-25.02	-11.69	-0.72
Gathering of wild fruits and berries, preservation:			
– small organizations	29.78	7.23	5.07
– medium-sized organizations	-5.43	4.18	3.13
Meat and milk production:			
– small organizations	-22.15	-16.92	-3.45
– medium-sized organizations	-3.39	0.02	0.01

The geopolitical factor influenced the decrease in the number of innovators with international organic certificates. In 2023, the number of organizations certified according to international organic standards was only 59; (note, in 2020 there were 75 such organizations). The map shows organizations certified according to international organic standards in 2023 (see Fig. 3).

Innovators of organic agriculture continue to remain within the borders of Moscow, the Moscow and Kaluga regions. Types of activity are: (1) mixed agriculture, (2) production of dairy food products. Main organizations are: (1) «Savinskaya Niva» LLC, (2) «Bogimovsky Cheese Factories» LLC, (3) «Shulgino» LLC. There are still related industries in Moscow (processing and export of products). From 2023, the Moscow region became more focused on domestic consumers. The second center of concentration of organizations certified according to international organic standards remains the Tomsk region. However, the Tomsk region is the territory of only one innovator – «TDS-Group» LLC and its subsidiaries [7]. The possibilities for the transition of organic agriculture to the «densification» stage are determined by economic factors. In the short term, the most relevant factors are the demand for organic products.

3.2 Demand Factors for Organic Products

Price for organic and non-organic products. The most common organic products on the Russian market are vegetables and fruits, meat and dairy products [8]. The National Statistical Office does not provide information on organic products, so we propose to use

Fig. 3. Placement of organizations certified according to international organic standards in 2023

the international experience and data from the USDA Economics, Statistics and Market Information System (ESMIS) to compare prices, namely the USDA report on weekly advertised retail prices for fruits and vegetables [9]. One type of product from each product group was selected. For some types of organic products, prices are significantly higher (Fuji Apples + 30%, Tomatoes + 61%, Potatoes + 131%, Beef (fillet steak) + 260%, Yogurt (Greek) + 31%). Therefore, according to the income level of consumers, the food market can be divided into organic and inorganic segments.

Consumer income. Demand for organic products is directly dependent on consumer income. For the period 2019–2022, retail sales of organic products on the Russian market have stabilized [8]. According to the Russian Unified Interdepartmental Information and Statistical System (EMISS), in Russia over the past few years the purchasing power of the population's cash income has not significantly changed. All this does not contribute to increasing demand for organic products.

Consumer preferences. Promoting a healthy lifestyle has a positive impact on consumer demand. Consumers are beginning to care about their own health and are gradually changing their preferences [10, 11]. In Russia, promotion of the consumption of organic products comes mainly from public associations, one of which is the Union of Organic Farming. The Union conducts marketing research, training in organic agriculture, and publishes materials in the media [5].

Number of consumers. In section «Consumer income» we note that retail sales in the domestic market are not increasing and this may be due to a decrease in the number of consumers. In the Central Federal District, an increase in the permanent population is observed only in Moscow and the Moscow region. The main consumer demand for organic products has formed in these regions. This is facilitated by high per capita incomes and consumer proximity to the production of organic products (most organic farms are located in the Moscow region and adjacent regions).

Since 2022, organic products have substitute products with a lower price (due to lower costs). On March 1, 2022, Federal Law No. 159-FZ of June 11, 2021 «On Agricultural Products, Raw Materials and Food with Improved Characteristics» came into force. The Law introduced the concepts of «improved agricultural products» and «improved food». For these products, labeling is provided in the form of a graphic image, and a «register of manufacturers of agricultural products, food, industrial and other products with improved characteristics» has been created. For the consumer, the emergence of the concept of «improved products» may become confusingly similar to the concept of «organic products». Even if this is not happened, the producers of «organic» and «improved» would target the same target market segment. This may lead some consumers to choose improved products. In this case, there can be a decrease in consumer demand in the Russian market of organic products.

Consumer expectations. When purchasing organic products, consumers expect positive health benefits [12]. Another reason to buy organic products may be their taste. Research shows that health factors significantly influence consumers' willingness to buy organic products [13]. Thus, consumers may view organic purchases as an investment in their health [14].

4 Conclusions

Therefore, we have observed segmentation in the domestic food market. The identification of the organic food segment in the Russian market is determined by the desires and capabilities of consumers. The trends in increasing demand and supply in the organic food market are uniform. It can be assumed that the limiting factor of demand in the national market is the income of the population. Moscow and the Moscow region are traditionally classified as the regions with a high level of per capita income and an increasing population. By extrapolating trends in the location of producers and retail sales of organic products, we can assume that these markets will retain their leading positions. Regional markets in the eastern part of the country do not have these characteristics. At the same time, in regional markets, factors constraining supply will remain an undeveloped distribution system, lack of experience in the production of organic products and the expectation of low financial results.

In 2020–2023, organic agriculture developed mainly according by inertia, the total number of organic farmers is changed slightly. We observed structural changes: a decrease in the share of farmers with international certificates and an increase in the share of farmers certified according to national standards. In the third stage, «densification», there should be an increase in the number of followers throughout the territory.

Taking into account the presented spatial changes and institutional conditions, the scaling of organic agriculture and the transition to the «densification» stage will be determined by the possibilities of implementing government stimulating policies.

References

1. Defining innovation. https://www.oecd.org/site/innovationstrategy/defininginnovation.html. Accessed 21 July 2023
2. Padel, S.: Conversion to organic farming: a typical example of the diffusion of an innovation. Sociol. Rural. **41**(1), 40–61 (2001)
3. Galkin, D.: Spatial diffusion of organic agriculture: theoretical model and russian experience. In: Polyakov, R. (eds) Ecosystems Without Borders. EcoSystConfKlgtu 2021. Lecture Notes in Networks and Systems, vol 474, pp. 89–98. Springer, Cham (2022). https://doi.org/10.1007/978-3-031-05778-6_9
4. Hagerstrand, T: Innovation diffusion as a spatial process. In: Innovation Diffusion as a Spatial Process, University of Chicago Press (1968)
5. Union of Organic Farming. https://soz.bio/. Accessed 10 Aug 2023. (in Russian)
6. List-Org: Directory of organizations. https://www.list-org.com/. Accessed 11 Aug 2023. (in Russian)
7. Galkin, D.: Towards scalable consortium-based organic agri-food systems. In: Polyakov, R. (eds) Ecosystems Without Borders 2023. EcoSystConfKlgtu 2023. Lecture Notes in Networks and Systems, vol 705, pp. 112–122. Springer, Cham (2023). https://doi.org/10.1007/978-3-031-34329-2_12
8. FiBL: Data on Organic Agriculture in the World, Research Institute of Organic Agric. https://statistics.fibl.org/. Accessed 8 July 2023
9. USDA Economics, Statistics and Market Information System (ESMIS). https://usda.library.cornell.edu/concern/publications/. Accessed 12 June 2023
10. Schösler, H., Joop, D.B., Jan, J.B.: The organic food philosophy: a qualitative exploration of the practices, values, and beliefs of Dutch organic consumers within a cultural–historical frame. J. Agric. Environ. Ethics **26**, 439–460 (2013)
11. Paul, J., Jyoti, R.: Consumer behavior and purchase intention for organic food. J. Consum. Mark. **29**, 412–422 (2012)
12. Hansmann, R., Ivo, B., Claudia, R.B.: Increasing organic food consumption: an integrating model of drivers and barriers. J. Clean. Prod. **275**, 123058 (2020)
13. Gundala, R.R., Singh, A.: What motivates consumers to buy organic foods? Results of an empirical study in the United States. PLoS ONE **16**(9), e0257288 (2021)
14. Nagaraj, S.: Role of consumer health consciousness, food safety and attitude on organic food purchase in emerging market: a serial mediation model. J. Retail. Consum. Serv. **59**, 102423 (2021)

Corporate Environmental Responsibility: Exploring Companies' Efforts to Reduce Their Environmental Footprint and Develop Sustainable Business Practices

Elena Nikiforova[1] and Ruslan Polyakov[2(✉)]

[1] Financial University Under the Government of the Russian Federation, Moscow, Russia
[2] Kaliningrad State Technical University, Kaliningrad, Russia
ruslan.polyakov@klgtu.ru

Abstract. Corporate environmental responsibility plays one of the key roles in the modern world, where environmental issues are becoming an integral part of not only society, but also business. In the context of the current trend towards the transition of business to a "green" course and environmental orientation, it is important to track the vector of effective practices for business in terms of improving its sustainability and environmental responsibility. This article examines the activities of corporations that demonstrate a high level of environmental responsibility, as well as what problems exist in this area and how it can be improved. The article looks at the various strategies and practices that companies are adopting to reduce their environmental impact as well as develop sustainable business models. It also looks at the factors that favor the company's operations through environmental responsibility. Considering all the aspects that affect environmental conservation, business practices are developed, taking into account various systems and marketing moves which include: ESG systems, environmental compliance, eco-marketing, etc. The article enhances the understanding of the importance of environmental responsibility at the corporate level and contribute to the creation of a sustainable economy.

Keywords: ESG · Ecology · Responsibility · Corporations · Sustainability

1 Introduction

In today's world, environmental responsibility has become an integral part of any sphere of life, both society and corporations. The relevance and importance of environmental responsibility issues is becoming more and more serious every year, and the role of corporations in this context is one of the key ones. Factors such as global warming associated with increased greenhouse gas emissions, rising water levels, the use of non-environmental materials, social pressure, etc. exacerbate the need for action at the corporate level.

"Environmental responsibility is the obligation of a business to conduct its operations in a way that keeps the environment safe" [1].

R. Polyakov (Ed.): EcoSyst 2023, LNNS 1092, pp. 95–103, 2024.
https://doi.org/10.1007/978-3-031-67354-2_11

"A company's environmental responsibility takes into account economic, social and internal reasons for preserving and enhancing the environment, is linked to the company's long-term business models, and must comply with expected regulatory requirements" [2].

The environmental responsibility of corporations provides them with significant advantages such as: saving resources, minimizing energy costs, attracting new customers, investors, etc. The role of innovation and technology used by corporations to improve the environmental situation is also important. It is important to note the role of innovations and technologies used by corporations to improve the environmental situation. The use of environmentally friendly production methods and technologies provides new opportunities for corporations to reduce their ecological footprint, and also affects their competitiveness and image.

As in foreign countries, so in the Russian Federation, the sphere of environmental responsibility is regulated through the introduction of laws and regulations that require corporations to comply with environmental standards, and their violation can lead to fines. Supervision over the implementation of environmental legislation is carried out through environmental compliance, which is reduced to the construction of such a mechanism in the company, through which environmental and legal risks are minimized by working with many resources: material, human, information, etc..

"Environmental responsibility can be regarded as the highest level of manifestation of business awareness, when with sufficient compliance with mandatory environmental requirements established by law, it voluntarily implements environmental policy (a set of actions and measures to ensure the quality of the natural environment of the territory, which translate the basic principles and intentions of the organization regarding environmental protection), aimed at minimizing, compensating and preventing damage, environmental restoration" [3].

2 Materials and Methods

Environmental responsibility consists in the readiness of business "to voluntarily and consciously accept in all processes of economic activity additional (in excess of those fixed by law) environmental obligations through the development and implementation of environmental policy based on the principles of sustainable development" [3].

This article examines the experience of both foreign and domestic corporations. It analyzes positive and negative experience in terms of environmental responsibility, what programs are used to reduce the ecological footprint, as well as how motivation to preserve nature, customers and employees takes place.

Theoretical and practical aspects of environmental responsibility of business are considered in the works by A. M. Andreeva. [3], Larionov V. G. [4], Bondarenko Yu. [5], Dubovik D. M. [6], Kvanina V. V. [7], Petrov A. M. [8], Efimova O. V. [9], Stepanova T.E. [10], Afanasyev M.V. and Shash N.N. [11] and others.

In order to achieve the research objectives, this paper uses methods of content analysis of scientific publications and materials of various corporations with subsequent evaluation and systematization of data.

3 Discussion

In the current realities, environmental responsibility is popularized, affecting all spheres of life, both society and business. Most large companies realize that not only economic benefits but also commitment to environmental standards are important for their long-term sustainability and image. Growing environmental challenges and societal pressures require a shift towards more responsible environmental practices.

Table 1. Eco-initiatives of foreign companies [12, 13].

Company	Country	Programs
ExxonMobil	U.S	Introduces innovative IT solutions: use of cloud technologies in terms of data collection and analysis; Project Astra provides control of methane leakage, using specialized sensors
Vale S.A	Brazil	Uses technology for autonomous driving. This technology reduces carbon dioxide emissions. Optimization of logistics ensures the efficiency of mining operations and also extends the life of the equipment
Georgia-Pacific and Resolute Forest Products	USA and Canada	They use environmentally friendly fuels in their operations - natural gas and biomass
Chevron	U.S	Partnered with California Bioenergy LL to convert methane produced by California dairy farms into renewable natural gas C
Nippon Steel Corporation	Japan	Apply technologies to recover waste heat and by-product gas from steelmaking to power their own operations
Body Shop (Natura & Co.)	UK	In the late '70s, suggested that consumers purchase cosmetics in reusable packaging, and refused to test cosmetics on animals
Nike	U.S	Launched a program to build new playgrounds out of recycled sneakers
Ben & Jerry's (Unilever)	UK	The company supplemented its financial report with an overview of its environmental impact. The practice of corporate social responsibility reports is now issued by major companies
McDonald's	U.S	Reduces carbon footprint in meat and packaging production
Netflix	U.S	Reduces energy consumption and invests in restoring ecosystems that naturally store carbon dioxide

"The modern world trend consists in the formation of environmentally oriented business, which being one of the types of sustainability-oriented development realizes the transformation of consumer negative value into consumer value" [4].

Let us take a closer look at the eco-initiatives of foreign companies (Table 1).

Regardless of the industry, many large companies are going green and minimizing their negative footprint, introducing new methods and solutions to problems that resonate with other large corporations as well.

There are many domestic companies in the Russian Federation that follow environmental responsibility and are also a source of motivation not only for employees but also for customers (Table 2).

Despite a sufficient number of corporations with a high level of environmental responsibility, in Russia today the level of pollution from large companies still remains at a high level. For example, Russian Railways accounts for more than half of its cargo transportation by carbon cargo - coal and oil products,

It should be noted that "both foreign and domestic experience shows that not always the producers of products, calling themselves producers of environmentally safe products, are such in fact, and the products they produce are ecological. This discrepancy is called "greenwashing" [16]. In Russia, this term is known as green camouflage.

Greenwashing is "the environmental positioning of a company, product or service that is not supported by any evidence" [17].

Many companies "camouflage" their negative impact on the environment. Such a marketing move is caused by the reluctance of companies to fully engage in environmental activities and bear responsibility for them, but modern realities oblige to be involved in the preservation of the environment.

4 Research Results

In today's world, the environmental responsibility of corporations is not always the case. There are still many problems that need to be solved, and it is necessary to develop control at the state level in terms of environmental legislation. As we mentioned earlier, one of the effective ways of control is environmental compliance.

Let us consider in more detail several definitions of this term (Table 3).

Environmental compliance has several objectives (Fig. 1).

The term environmental compliance has not yet been regulated in Russian law, and therefore it is not appropriate to rely on preferences. However, "despite the lack of legislative regulation, changes in the government's environmental policy are already emerging, creating the ground for future encouragement of compliance. From May 1 of this year until December 31, 2023, Rosprirodnadzor is conducting an experiment on consulting investment projects at the stage of creation in order to bring them into compliance with environmental legislation" [11].

Along with environmental compliance to regulate the environmental activities of corporations can be put ESG-transformation. "Within the framework of the established approach, the concept of ESG is considered either in the context of transition to green growth - economic growth that ensures the maintenance of biodiversity, conservation and rational use of natural resources and the environment, reduction of negative climate

Table 2. Eco-initiatives of domestic companies [14, 15].

Company	Programs
Russian division of Mars Corporation	Since 2015, Russian factories have not taken waste to landfills, incinerating only about 10%. The rest is recycled or put to beneficial use, for example, as feed for productive animals
Gazprom Neft	The Corporation has introduced a number of initiatives: – reduction of technogenic load and maintenance of favorable condition of the natural environment; – avoidance of environmental damage from economic activities; – implementation of the world's best practices in the field of environmental safety; – rational use, restoration and protection of natural resources; preservation of biodiversity in the regions where we operate
Sberbank	Has a policy to reduce the consumption of paper, energy and fuel for its own cars, is engaged in landscaping of the territory
X5 Retail Group	Approved a sustainable development strategy in terms of reducing greenhouse gas emissions, increasing the share of recyclable food products and the share of own brands in eco-friendly packaging
Splat	The biodegradability of the products exceeds 80% and they do not contain substances hazardous to humans and the environment. Supports CO2-neutrality program - the company together with its partners planted 45.2 hectares of trees. From 2011 to 2019, the company reduced specific water consumption by 32% and energy consumption by 23%
"Rosatom"	Participates in the disposal of nuclear submarines and builds modern radioactive waste storage facilities and wind generation parks, and has begun to develop a system for managing Class I and II hazardous waste
"Alrosa"	In 2019, the company allocated RUB 7.7 billion, or about 3.2% of its revenue, for environmental purposes, which is one of the highest figures among Russian mining companies. The company spent more than half of this amount on waste management (81.8 million tons in total). The share of recycled and reused water supply at the group's enterprises at the end of 2019 is 55% of the total water intake, which amounted to 7.6 million cubic meters

change, or in terms of compliance with social responsibility requirements, both in the external and internal circuits" [19].

Table 3. Definitions of environmental compliance.

Company	Programs
Author	Determination
Y. Bondarenko	"An integral part of a company's corporate culture in which the performance of each employee's job responsibilities, including decision-making at all levels, must meet the standards of legality and integrity set by the company for the conduct of its business" [5]
T.N. Mikhnenko	"A system of mechanisms and levers by which the likelihood of violating the law in general is reduced" [18]
D.M. Dubovik	"An independent direction in the system of corporate governance of the enterprise-natural resource user, which generally involves independent control over compliance with environmental legislation in order to minimize legal and reputational risks" [6]
V.V. Kvanina, T.I. Makarova	"A program that includes the optimal set of conservation measures required in a given case, for a given type of economic activity, in order to implement the conservation policy of a particular entity" [7]

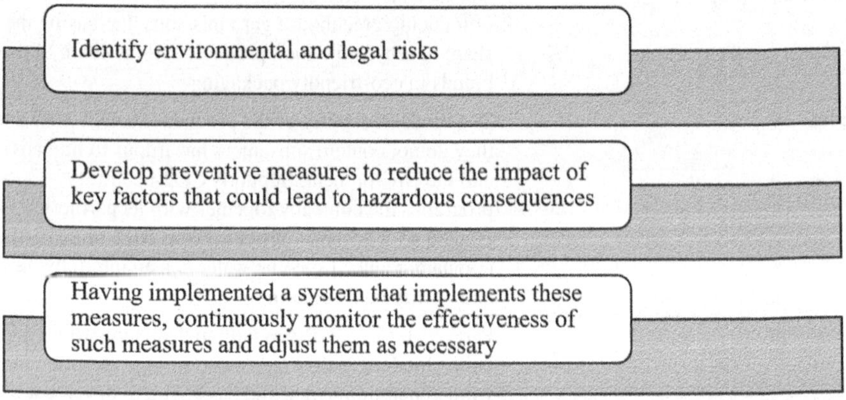

Fig. 1. Objectives of environmental compliance.

The essence of an ESG system includes the following basic principles that companies must comply with:

- environmental (environmental). Reduction of negative impact on the environment, preservation of biodiversity, environmental transparency in terms of documentation;
- social (social). Observance of human rights, responsibility to society, observance of labor rights of the company's employees, etc.;
- governance. Transparency of reporting, compliance with legislation, ensuring oversight of the company's activities.

The application of ESG and environmental compliance principles represent fundamental values and norms that help companies to achieve sustainable development as well as to take into account the interests of different stakeholders.

The development of sustainable environmental business practices by companies should follow certain principles and steps (Fig. 2).

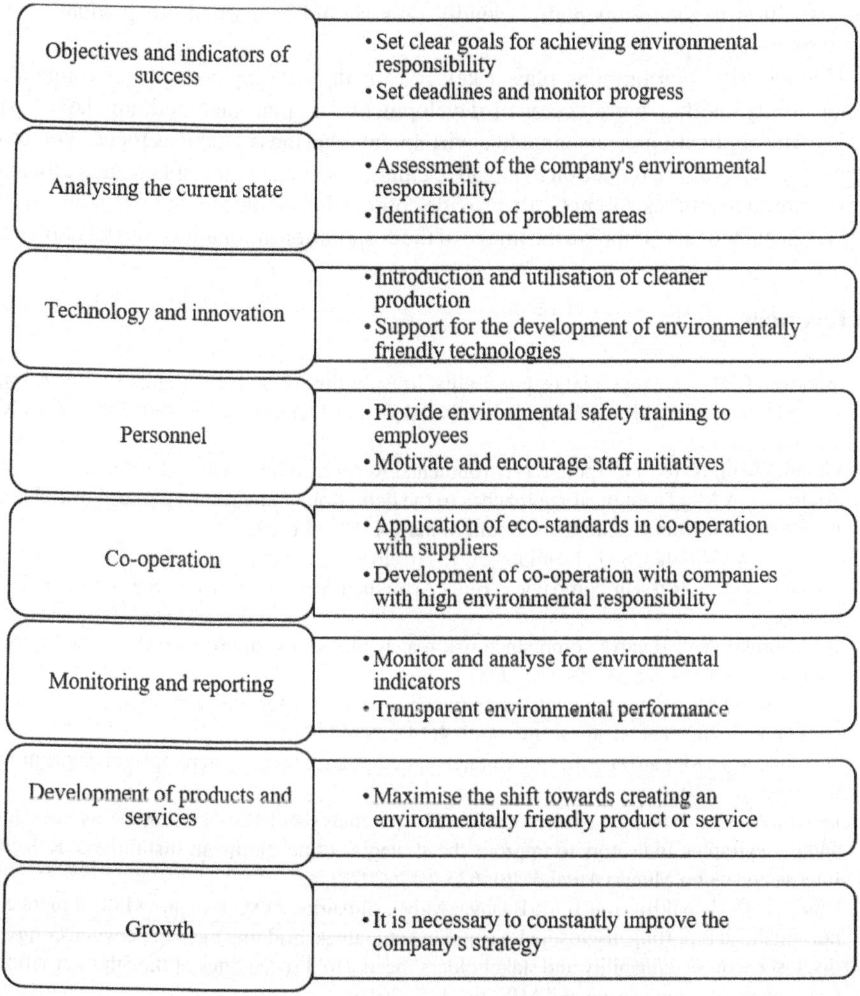

Fig. 2. Objectives of environmental compliance.

The role of sustainable environmental business practices for corporations is quite significant, for several reasons, as the conditioned:

- reducing the ecological footprint;
- economic benefit;
- Improving reputation and image;

– team building.

5 Conclusion

Analyzing all of the above, it should be noted that the situation with environmental responsibility in Russia is quite challenging and requires a large amount of work. The transition to environmentally friendly rails should be carried out gradually and as productively as possible.

Undoubtedly, corporations play a key role in the development of environmental responsibility, as they are a vector of development for small and medium-sized businesses. The application of sustainable environmental business practices focused on biodiversity conservation, reduction of negative emissions into the atmosphere, and efficient use of natural resources will not only provide corporations with economic benefits in the long term, but will also improve the image of the corporation among various stakeholders.

References

1. Windsor, D.: Corporate social responsibility, irresponsibility, and citizenship. In: Encyclopedia of Organizational Knowledge, Administration, and Technology, pp. 830–846 (2021). IGI Global
2. Gunningham, N. (ed.) Corporate Environmental Responsibility (2017). Routledge
3. Andreeva, A.M.: Theoretical approaches to the definition of environmental responsibility of business. Bull. Buryat State Univ. Econ. Manag. **4**, 10–17 (2022)
4. Larionov, V.G., Falko, S.G., Demidov, A.V.: An environmentally oriented model of integrated management of Russian companies. Bull. Astrakhan State Tech. Univ. Ser. Econ. **4**, 7–18 (2018)
5. Bondarenko, Y.: Effective compliance risk management: a systematic approach and critical analysis. Corp. Lawyer **6**, 29–32 (2008)
6. Dubovik, D.M.: Environmental compliance as a tool for the development of green entrepreneurship. Theor. Appl. Jurisprud. **4**(14), 39–44 (2022)
7. Kvanina, V.V., Makarova, T.I.: Environmental compliance in the system of legal environmental protection (2020)
8. Petrov, A., Churilova, E., Nikiforova, E.: Cluster analysis of Russian regions by economic and demographic indicators to improve the strategies of development institutions. Relações Internacionais no Mundo Atual **3**(36), 623–647 (2022)
9. Efimova, O.V., Nikiforova, E.V., Basova, M.M., Shnaider, O.V., Ushanov, I.G.: Practice of non-financial reporting disclosure by Russian companies: bridging the gap between company disclosures on sustainability and stakeholders' needs. In: Proceedings of the 5th International Conference on Engineering and MIS, pp. 1–5 (2019)
10. Stepanova, T., Schneider, O.: The influence of social indicators of economic actors on the construction of an effective ecosystem. In: Polyakov, R. (ed.) Ecosystems Without Borders: Opportunities and Challenges, pp. 9–18. Springer International Publishing, Cham (2022). https://doi.org/10.1007/978-3-031-05778-6_2
11. Afanasyev, M.P., Shash, N.N.: ESG transformation in the corporate sector: systematization of a global approach. Forecast. Prob. **6**(195), 185–197 (2022)
12. Shamsadov, M.S.: Ecological footprint in marketing. Colloquium J. **15**(2), pp. 9–11 (2021). Golopristanskiy Regional Employment Center

13. Most environmentally friendly companies in Russia, forbes rating. https://xakac.info/news/94498. Accessed 6 Feb 2024
14. Mikhaleva, A.A., Subkhonberdiev, A.: Technological sovereignty as a strategy for the future development of Russia within the framework of the economic decarbonization policy. In: Economic Aspects of Environmental Management: Traditions and Innovations, pp. 115–118 (2023)
15. Skaranik, S.S.: Environmental responsibility in modern corporate governance practices of Russian companies. Econ. Constr. Environ. Manag. **2**(75), 15–23 (2020)
16. Voronina, N.P.: Greenwashing: a legal mechanism of counteraction. Bull. Univ. OE Kutafin, **3**(103), 28–36 (2023)
17. Finogentova, A.V.: To the concept of greenwashing. In: A Healthy Environment is the Basis for Regional Security: Materials of the First International Environmental Forum in Ryazan, pp. 365–368 (2017)
18. Mikhnenko, T.N.: The impact of antimonopoly compliance on the efficiency of functioning of business entities and the economy as a whole. World Sci. Pedagogy Psychol. **2**, 36 (2015)
19. Pozdnyakova, P.: Environmental compliance: how systemic instruments reduce environmental and legal risks of enterprises. https://journal.ecostandardgroup.ru/eco/praktikum/ekologicheskiy-komplaens-kak-sistemnye-instrumenty-snizhayut-ekologo-pravovye-riski-predpriyatiy/

Green Infrastructure and Urban Planning: Shaping Sustainable Cities of the Future

Tatyana Stepanova[1](✉) and Olga Schneider[2]

[1] Kaliningrad State Technical University, Moscow, Russia
tatyana.stepanova@klgtu.ru
[2] Financial University Under the Government of the Russian Federation, Moscow, Russia

Abstract. Today the leader in the urban planning agenda is the vector for the creation of sustainable cities of the future. In this vein, analyzing the sustainable development of cities with developed green infrastructure seems appropriate and relevant. This paper investigates the impact of green infrastructure and urban planning on the environment and the well-being of the city's population. It also describes the existing ecosystem services and what kind of legislative support is provided in terms of ecology. In today's world, green infrastructure is only gaining popularity, which is undoubtedly important in the context of global environmental problems. Green infrastructure and urban planning are an integral part on the way to a cleaner and healthier urban policy, minimize the negative impact, motivate the state, business and society as a whole on the path of environmental awareness. The need for cooperation of all stakeholders ensures the preservation of biodiversity, helps to adapt to the climate, improves the health of both nature and society, and reduces the negative impact on the environment. Understanding, adopting and implementing modern green initiatives in urban planning will ensure a sustainable future for modern cities.

Keywords: Green Infrastructure · Urban Planning · Shaping Sustainable Cities

1 Introduction

Green infrastructure and urban planning are integral components of urbanization in today's world. Shaping these aspects of the future is part of a sustainable development strategy that takes into account such factors as: economic, social, environmental and qualitative. Cities with developed infrastructure and all aspects of comfortable life, from the arrangement of city streets to a unified design code, are considered more sustainable, so the role of green infrastructure and urban planning occupy one of the key roles in the formation of sustainable cities of the future.

Green infrastructure is a network of natural and semi-natural areas that are planned and maintained to provide multiple benefits to people [1].

Green infrastructure provides ecosystem services, which include all the benefits that people get for free from natural resources, e.g.: greening of urban areas ensures that air pollution is cleared from the air, as well as adaptation to climate change through

R. Polyakov (Ed.): EcoSyst 2023, LNNS 1092, pp. 104–110, 2024.
https://doi.org/10.1007/978-3-031-67354-2_12

natural resources. The role of insects and birds, as well as soil, etc., is not insignificant in this context. Thus, this infrastructure has not only an aesthetic function, but is also an effective tool for solving many urban issues.

It is worth noting that in addition to the above resources, green infrastructure and urban planning provide cities with economic benefits in the form of additional investment in the city, job creation and increased attractiveness of the city.

The article looks at the main aspects of green infrastructure, the interaction with urban planning, what challenges exist in implementing this system and how all this will influence the creation of sustainable cities of the future.

2 Methods and Materials

Green infrastructure is "an interconnected system of different green spaces, from 'wild' natural areas to 'cultivated' areas such as suburban agricultural areas, urban parks, public gardens, etc. Assumes a continuous connection between them and unimpeded movement in it of various species of animals and plants inherent to the area, and is designed to favorably affect the quality of life of the population" [2].

Theoretical and practical aspects of building green infrastructure and sustainable urban planning are considered in the works by Podoynitsina D.S. [2], A. I. Kotov, A. Likhtin. A. [3], Kudryavtseva O.V., Khovavko I.Yu. [4] and others. The study also used as input data the materials of the United Nations and the state development corporation VEB.RF, as well as scientific works in the field of sustainable development, urbanism and green infrastructure. Achievement of the objectives of this study was carried out using the method of content analysis of scientific publications and materials with their subsequent study, evaluation and systematization of data.

3 Discussion

Conservation of nature, reduction of negative footprint, formation of green economy and infrastructure - all these are acute topics in the modern world. Despite the global trend and trends in environmental awareness, the negative impact on the environment remains at the same level. Human activities have changed almost 75% of the Earth's surface, pushing animal life and nature to a very small corner of the planet. According to the 2019 Global Biodiversity and Ecosystem Services Assessment Report, about one million animal and plant species are at risk of extinction, and many of them are at risk of extinction within the coming decades [5].

Green infrastructure represents such natural elements as: parks, forests, rivers, lakes, squares, etc. they are integrated and designed taking into account ecological and sustainable principles in order to improve the quality of urban life and form sustainable cities. Also, it fulfills certain functions to ensure comfortable life (Fig. 1).

Ecosystem functions in the current realities balance urban growth and conservation of natural resources.

The UN Millennium Ecosystem Assessment methodology [6] divides such services into 4 classes (Fig. 2) [4] divides such services into 4 classes (Fig. 2).

Ecosystem functions	Air and water purification
	Climate control
	Conservation and maintenance of biodiversity
	Recreational spaces
	Adaptation to climate change

Fig. 1. Green infrastructure functions.

Supporting ecosystem services

•Natural processes required to maintain other ecosystem services, e.g. soil formation, water and nutrient cycling, photosynthesis

Regulating ecosystem services

•Benefits derived from regulating ecosystem processes. These include regulating climate processes, controlling temperature, reducing the negative effects of natural disasters, regulating erosion, cleaning water, wastewater and air, pollination, etc., and regulating ecosystem services

Provisioning ecosystem services

•The products that people obtain from ecosystems. These include fuel, crops, harvested animals, seafood, food from plants, animals and microbes; materials including timber, cotton, wool, silk, etc.; drinking water, drinking water, drinking water, etc. Drinking water, water used for irrigation and industry, etc., etc.

Cultural ecosystem services

•Cultural, educational and spiritual benefits people receive from ecosystems. These include cultural, spiritual and religious diversity and development; recreational and ecotourism opportunities; scientific research, education; aesthetic values due to the different properties of ecosystems

Fig. 2. 4 classes of ecosystem services of the UN methodology (compiled by the authors based on [6]).

"Ecosystem services" is a deliberately anthropocentric phrase that forms an argument for the conservation of natural systems in the terminology of economics. This language and demonstration of economic value leads to greater fixation and investment in their preservation" [7].

Ecosystem services are one of the components of urban planning and play an important role in shaping the cities of the future.

Urban planning plays an important role in the provision of green infrastructure, which seeks to create sustainable urban environments where quality of life and sustainable development are the main criteria.

The increased activity of the population and their desire to participate in decision-making related to their living environment is based on the principles of the Charter of New Urbanism, adopted in 1999, which proclaimed the main task of urban planners to create cities comfortable for human habitation [3]. The principles of the Charter of New Urbanism are shown in Fig. 3.

Both green infrastructure and urban planning are closely linked and share the same goal of creating sustainable cities of the future. They address many aspects of human and urban life, such as: climate change, respect for nature, sustainable urbanization,

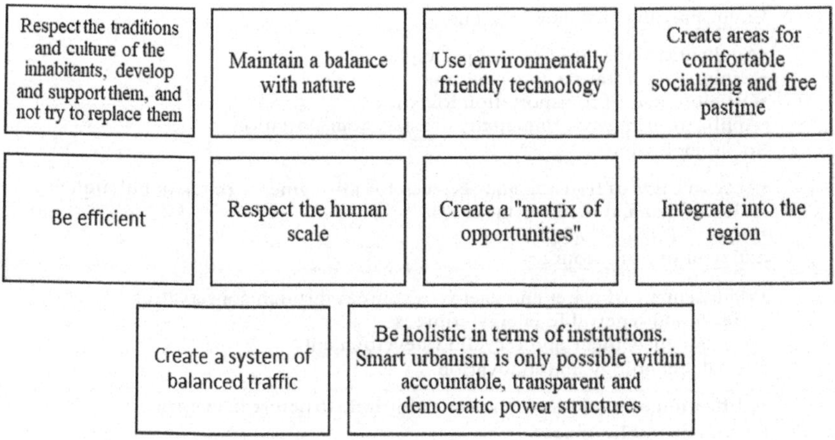

Fig. 3. Principles of the Charter of New Urbanism (compiled by the authors on the basis of [8]).

increased mobility, and so on. By working together, they are creating cities that are sustainable, healthy and livable places, which is undoubtedly a key challenge in times of urbanization and natural change.

4 Research Results

The formation of sustainable cities of the future in today's world is only in its initial stages and requires some effort. Sustainable urban development implies "…ensuring safety and favorable conditions for human life in the implementation of urban planning activities, limiting the negative impact of economic and other activities on the environment and ensuring the protection and rational use of natural resources in the interests of present and future generations" [9].

It should be noted that the main attributes inherent in a sustainable city are not devoid of specificity. Thus, most modern urbanists "a sustainable city is characterized by a favorable environmental situation, pedestrian accessibility of important objects, convenient logistics, housing of various types, high-quality urban environment, high quality of life, efficient use of space, healthy social climate, the presence of communication with the population of the city and taking into account its interests" [10].

Sustainable cities of the future must meet the social, environmental and economic aspects to ensure a high quality of life for the population, as well as provide an appropriate level of environmental protection. Let us consider some key aspects of the formation of such cities (Fig. 4).

In order to create this kind of cities, the support of the state, business and population is necessary. In terms of state regulation in Russia, there is the Charter of Sustainable Cities (ESG Charter) - a document describing the fundamental principles and key objectives of sustainable urban development as an element of comprehensive ESG transformation in the Russian Federation [10]. The principles of the Charter echo the principles of the Charter of New Urbanism:

Environmental sustainable planning

- Application of energy saving principles
- Minimizing negative emissionsсов
- Development of transportation logistics
- Application of environmentally friendly transportation

Social inclusion

- Accessibility of housing and services for all segments of the population
- Educational and medical facilities
- Support for cultural diversity

Efficient use of resources

- Management of water and energy resources through conservation systems and renewable energy sources
- Waste management and recycling development

Digital technology and innovation

- Utilization of modern technologies for infrastructure management

Green and public spaces

- Planning of parks and urban recreation areas
- Urban Greening

Adaptation to climate change

- Strengthening infrastructure in case of natural disasters
- Reducing carbon in the atmosphere

Community Involvement

- Motivating citizens to conserve and carefully use natural resources
- Support and development of ecologically oriented societie

Fig. 4. Key aspects of shaping sustainable cities of the future.

1.Principle #1: Environmental friendliness. The cities joining this Charter express their willingness to reduce negative environmental impact through the rational use of natural resources and minimization of negative emissions into the atmosphere. The basic paradigm of this principle is “prevention is better than cure” and the main objective is to recognize that environmental stewardship is everyone’s responsibility.

2.Principle #2: Diversity and Equality. Human rights and freedoms are recognized as the highest value. Observance and protection of the rights of city residents, respectful attitude towards human beings in society is a direct responsibility of every city. This value is also understood as social equality of the population, where there is no room for disadvantage on any grounds, be it gender, race, age or religion. Cities take all steps to ensure that the realization of these guarantees is fully enforced, respected and regulated. Cities recognize the need to apply practical measures for violations of human rights and freedoms.

3.Principle #3: High standards in management. Ensuring high standards and development of the urban environment directly depends on the interaction of the state, business, and society. The main goal of this principle is to ensure continuous cooperation of all stakeholders in unlocking the potential of the city, as well as creating a transparent management system.

Financial transparency of environmental activities of all structures is a guarantee of successful and effective development not only of cities, but also of the state.

The key intentions of the Charter participants are depicted in Fig. 5.

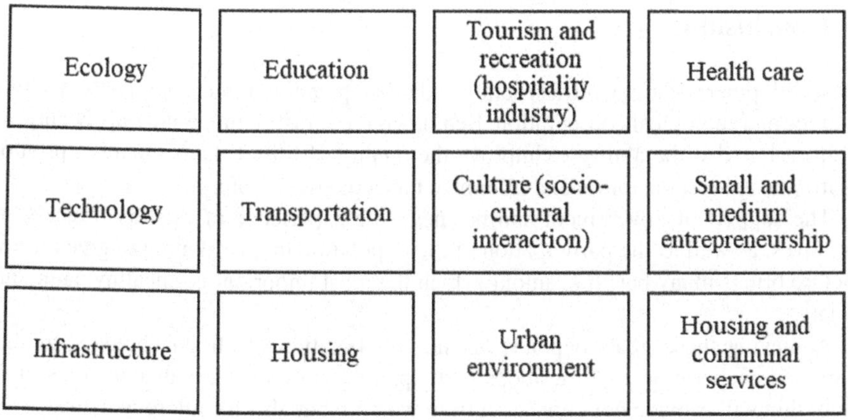

Fig. 5. Key intentions of the Charter participants (compiled by the authors on the basis of [11]).

Also, the state creates various projects in the field of urban and environmental improvement: "Housing and Urban Environment", "Formation of Comfortable Urban Environment", etc.

There are many benefits for businesses from participating in urban planning and green infrastructure. For example, the improvement of neighborhoods and parks is not only an investment in future development, but also affects the reputation of the company as a whole. Small and medium-sized businesses that operate in neighborhood buildings are also an economic benefit for businesses. At the moment there are various projects aimed at green business: WWF's "Living Office" project, whose participants are Russian companies, Green Office project by Greenpeace-Russia, Eco-Office project by Green Up agency, "Leaf of Life. Eco-office", etc.

In terms of society, there are many eco-organizations, bloggers, and activists in today's world with one single goal - to preserve the environment and motivate society to be environmentally responsible. For example, they create eco-friendly pieces of furniture, use biodegradable bags and utensils, promote the idea of sorting garbage, hold environmental festivals and popularize environmental awareness through media and social networks.

"In the sphere of eco-oriented communications, direct dialog between the authorities and the population, frankness and mutual trust of the parties, desire for cooperation are of paramount importance. Therefore, eco-oriented communications represent a significant potential and an effective means (tool) for the development and implementation of models of urban development, its management from the perspective of environmental improvement, the formation of patriotism, activity and responsibility of the population in relation to their city, participation in solving its problems. It is precisely because eco-oriented communications are fundamentally public - they affect the interests of all segments of the population, regardless of status, nationality, income level, etc., that these communications are a powerful basis for integrating the interests of the population. - these communications are a powerful basis for the integration of the interests of the city's population, the formation of the status and self-consciousness of the population as its subject" [11].

5 Conclusion

The development of green infrastructure and urban planning is an integral part of natural resource and ecological conservation. Behind green infrastructure is not only beauty and comfort, but also the ability to improve the quality of life of the population, promote environmental responsibility and minimize the ecological footprint.

The support of government and business, the experience of existing projects and initiatives, as well as the participation of the population in contributing to green infrastructure brings many benefits, among which the most important is a healthy nature and society.

Achieving these goals depends on the collaboration of all stakeholders, as their continued cooperation and conscious participation will create sustainable cities of the future that will provide not only for current worship but also for future worship.

Green infrastructure and urban planning is not just a trendy movement, but a strategic necessity for the cities of the future.

References

1. Naumann S., Davis M., Kaphengst T., Pieterse M., Rayment M.: Design, Implementation and Cost Elements of Green Infrastructure Projects, Final Report to the European Commission, DG Environment, Ecologic Institute and GHK Consulting (2011). 138 pages
2. LNCS Homepage. http://www.springer.com/lncs. Accessed 21 Nov 2016
3. Kotov, A.I., Likhtin, A.A.: Urbanistics and urban planning: symbiosis of knowledge and practice at the service of man. Manag. Consult. **2**(170), 81–89 (2023)
4. Kudryavtseva, O.V., Khovavko, I.Y.: In search of sustainable development: a critical analysis of the experience of modern Moscow experience. ENSR **2**(97) (2022)
5. UN Sustainable Development Goals. https://www.un.org/sustainabledevelopment/ru/sustainable-%20development-goals/. Accessed 17 Sep 2023
6. The Millennium Ecosystem Assessment (MEA) Program was launched in 2000 by UN Secretary-General Kofi Annan.
7. Austin G.: Green Infrastructure for Landscape Planning. Integration of Human and Natural Systems, Routledge, London and New York, 266 (2014)
8. Charter of New Urbanism. https://www.cnu.org/who-we-are/charter-new-urbanism. Accessed 15 Dec 2022
9. Town-planning code of the Russian Federation as amended on December 30, 2021 from 29.12.2004 N 190-FZ (2021)
10. VEB. RF: Charter of Sustainable Cities of Russia. https://вэб.рф/ustojchivoe-razvitie/khartiya-ustoi-chivykh-gorodov-rossii. Accessed 17 Sep 2023
11. Grigoryan, A.E.: Center and periphery in the space of eco-oriented communications of the city. In: Law and Practice, vol. 3, pp. 203–207 (2017). https://cyberleninka.ru/article/n/tsentr-i-periferiya-v-prostranstve-ekoorientirovannyh-kommunikatsiy-goroda

Sustainable Development Goals Achievement in the Context of Regional Economic Security

Albert Mnatsakanyan and Marina Pobegaylo(✉)

Kaliningrad State Technical University, Kaliningrad, Russian Federation
marina.pobegajlo@klgtu.ru

Abstract. The paper considers the study of the relation between the sustainable development goals proclaimed by the UN and Russian Federation development priority directions at both the national and regional levels. Paper reveals the actions to be taken in order to improve methodological approaches to measuring the degree of sustainable development goals achievement using indicators of economic security. Using the example of the Kaliningrad region, an analysis of economic security level in terms of ESG concept focus area was carried out. The main results of the region economic, social and environmental projections development analysis are presented, current trends and existing threats to economic security are identified, and ways to strengthen the level of regional economic security in terms of moving towards achieving sustainable development goals and implementing priority directions of the Russian economy are outlined. The authors proved the goals continuity in a multi-level hierarchical system, the possibility of breaking the sustainable development goals of the international community to the national and regional levels. Ways to improve the information and analytical base and tools for its formation are proposed to improve the quality of management decision-making to neutralize identified threats to regional economic security that impede the achievement of sustainable development goals. As a result of the study the projection of economic development was identified as a strong and priority projection, and the projection of the region social and environmental development as a requiring to strengthen by direct and immediate decision-making.

Keywords: Sustainable Development Goals · Economic Security · ESG

1 Introduction

Sustainable development of territories, industries and individual economic entities is closely connected with issues of ensuring economic security. The Sustainable Development Goals (SDGs), proclaimed in 2015 during the 21st session of the Conference of the Parties to the United Nations Framework Convention on Climate Change, as well as specified in the declaration "Transforming our world: the 2030 Agenda for Sustainable Development" [1], are reflected in determining the main directions of the Russian Federation sustainable development. Their implementation affects various levels, taking into account both spatial and temporal aspects.

R. Polyakov (Ed.): EcoSyst 2023, LNNS 1092, pp. 111–120, 2024.
https://doi.org/10.1007/978-3-031-67354-2_13

In accordance with the requirements of measurability, it is necessary to clarify the selection of criteria and individual indicators for assessing the achievability of sustainable development goals. A multi-level hierarchy is being developed, which includes both the SDGs themselves and the goals of economic security; the achievement of SDGs is possible only in the case of continuity and the absence of conflict between them within the framework of mutual penetration at each level of the complex hierarchy of the ecosystem.

2 Materials and Methods

The study is based on the following methods: causal, designed to identify cause-and-effect relationships between the goals of sustainable development of the region's ecosystem and indicators of its economic security; analysis and synthesis, the use of which made it possible to assess the economic security of the region in three projections of its development. During the study, an indicative method was also used, with the help of which quantitative values of the analyzed parameters were obtained. The use of retrospective analysis made it possible to determine the dynamics of changes in the main indicators of the economic security of the region's ecosystem, to identify challenges and threats, and the comparison method made it possible to mutually link the goals of sustainable development and indicators of the economic security of the region.

The process of assessing SDGs degree of achievement is complex and requires measurement and comparability of achieved results from the perspective of various criteria in the projections of economic, social, environmental development (ESG concept). The interrelation of these three projections, within the framework of which the main goals must be achieved, is the basis for the sustainable development of any of the levels of a complex hierarchical structure. At the same time, economic security is a multi-level system in which five levels are clearly distinguished: international, national, regional, economic entities and individuals (households). SDGs determine the goals of economic security, largely repeating them and detailing them depending on the level of the ecosystem being studied. The breaking of SD goals into parts by hierarchy levels and their relationship with economic security goals is of particular interest, as it brings us closer to the possibility of making a quantitative assessment of their achievability both in terms of ensuring the economic security of a particular subsystem (level) and specific sustainable development goals.

We chose the regional level as an object of study, using the example of which proved the relationship and continuity of economic security with the main directions of sustainable development. At that level we analyzed the components (subsystems) of a single ecosystem of the region, taking into account indicators of economic security in the aspect of sustainable development.

A number of researchers on the issue of ensuring economic security interpret its essence as an opportunity for development, economic growth, and the implementation of a special economic course, while adding the characteristics of a "green economy" to economic indicators to their analysis [2–4].

Analysis of the state of the region's economy is complicated by the need to take into account another aspect of security - the ability of the system to withstand existing external and internal threats. To confirm this postulate, we present the definition of V.V. Krivorotov: "the economic security of the region is a set of measures aimed at sustainable, continuous development and improvement of the region's economy, including a mechanism to counter external and internal threats."

3 Results

Determination of the diagnosing approach ensuring region economic security is important [5], it involves analyzing the main areas of the economy taking into account three criteria designed to reflect the features of the study of economic security, which fully correspond and repeat the projections of sustainable development (Fig. 1).

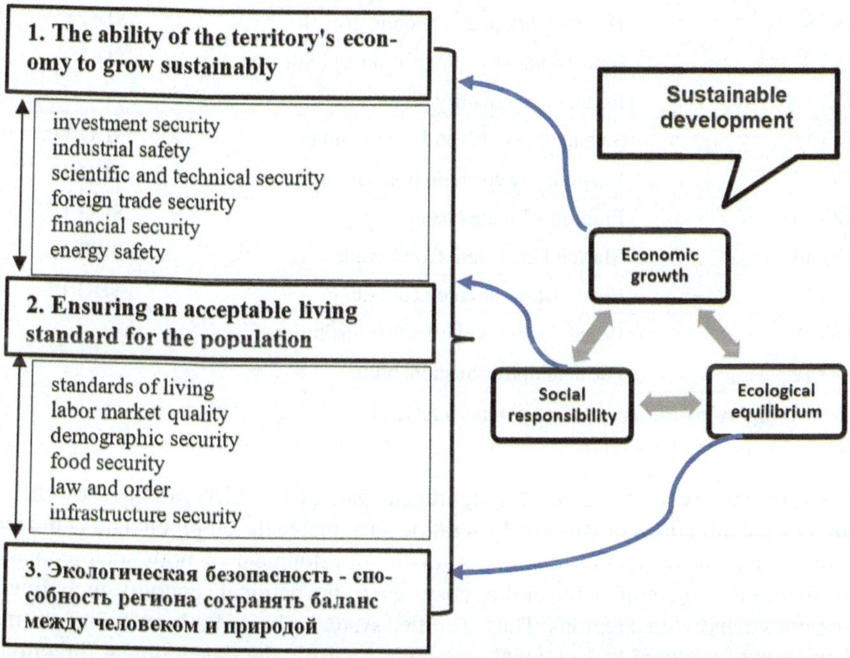

Fig. 1. The main directions for diagnosing the economic security of the region in terms of sustainable development xxSource: compiled by the authors using materials [4, 6].

Figure 1 clearly illustrates the interrelation of the region economic security components in the context of sustainable development projections, which proves the possibility of analyzing a broader concept of "sustainable development" term through a system of main indicators that simultaneously characterize both the state of ecosystem economic security and the results of specific SDGs achievement.

The details of the noted above fact are presented below in Table 1, which reflects the direct relationships between the SDGs defined by the UN and the goals and directions of sustainable development of the Russian Federation, approved by order of the Government of the Russian Federation dated July 14, 2021 [7].

Table 1. Relationship between the SDGs proclaimed by the UN and the SDGs of the Russian Federation

Symbol of the goal	UN SDGs	RF SDGs
SDG 1	Poverty eradication	
SDG 2	Ending hunger	
SDG 3	Good health and well-being	
SDG 4	Quality education	
SDG 5	Gender equality	
SDG 6	Clean water and sanitation	SDG 6
SDG 7	Inexpensive and clean energy	SDG 7
SDG 8	Decent work and economic growth	SDG 8
SDG 9	Industrialization, innovation and infrastructure	SDG 9
SDG 10	Reducing inequality	
SDG 11	Sustainable cities and communities	SDG 11
SDG 12	Responsible consumption and production	SDG 12
SDG 13	Fighting climate change	SDG 13
SDG 14	Marine Ecosystem Conservation	SDG 14
SDG 15	Preservation of land ecosystems	SDG 15
SDG 16	Peace, justice and effective institutions	
SDG 17	Partnership for Sustainability	

Source: compiled by the authors using materials[1, 7]

As we can see from Table 1, a significant part of the SDGs defined by the UN underlies the directions of Russian Federation sustainable development and is the basis for the formation of priority directions for the development of both socio-economic and environmental components of the ecosystem of the national economy, including the economies of individual regions. Thus, a unified system is being built, including the main subsystems of regional and national economic security, the development direction of which corresponds to the main goals and priority areas of Russian Federation sustainable development in unity with the SDGs approved by the UN.

Figure 2 below shows the interrelation and continuity of the goals and directions of sustainable development at different levels - from international to national, with the possibility of transformation to the regional level, industry level, the level of an economic entity and the individual (household).

The identity of the directions for assessing the effectiveness of achieving these goals both from the position of sustainable development (lower horizontal part of the figure - 3 aspects) and from the position of ensuring economic security (3 projections of the economic security system) allows us to use tools for assessing the degree of implementation of SDGs through existing methods used in the aspect of ensuring economic security, including individual indicators/indicators.

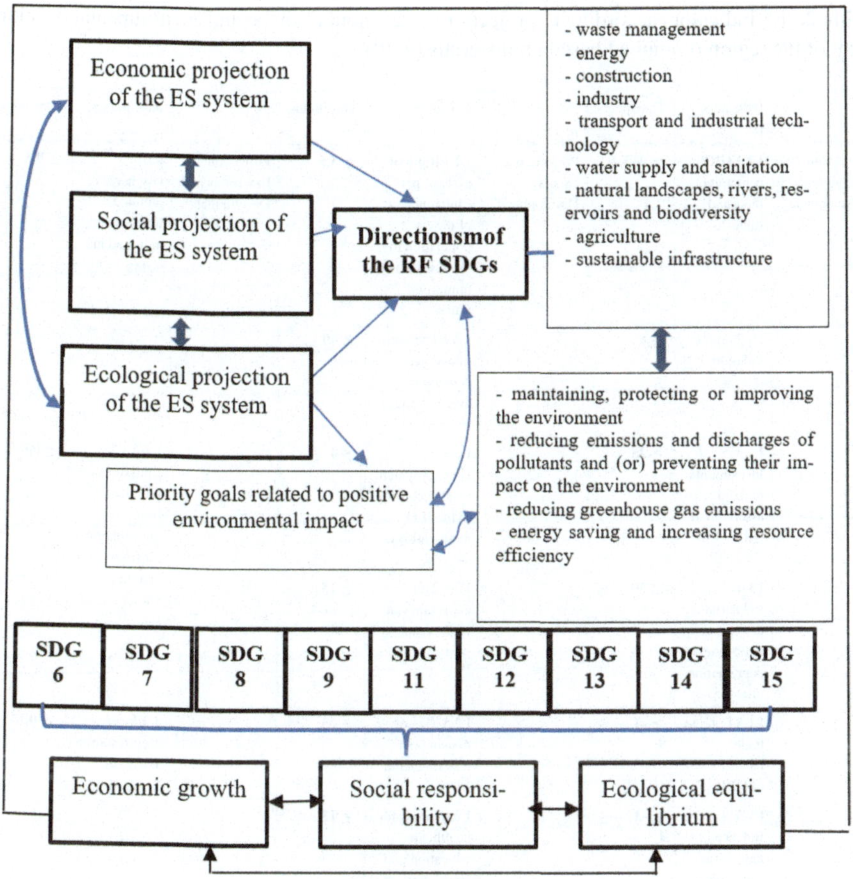

Fig. 2. Interrelation of SDGs and economic security projections Source: compiled by the authors using materials [1, 7].

A multi-level hierarchy in the case of assessing a regional cross-section of a unified system also demonstrates the comparability of the UN SDGs and the Russian Federation SDGs, which determines the possibility of measuring them in the aspect of ensuring the economic security of the region. In our opinion, the intersection of priority areas for sustainable development adopted in the Russian Federation not only contributes to achieving the SDGs listed at the bottom of the figure (SDG 6, 7, 8, 9, 11, 12, 13, 14, 15), but also corresponds and with other SDGs proclaimed by the UN. For example, the UN SDG 2 - eliminating hunger, is becoming achievable to a greater extent due to the development in the last decade of priority sectors that contribute to increasing the level of food security [8–10]. First of all, agriculture reflected in the official statistical reports which is confirmed by the results obtained.

Table 2 below lists the main indicators that form the basis of the tools for assessing the level of region economic security.

Table 2. ES indicators according to projections of economic, social and environmental development of the region (compiled based on materials [4, 6])

	Indicators (I)	Thresholds		Indicators	Thresholds		Indicators	Thresholds
Projection 1 "Economic Development"	I 1.1 GRP per capita, thousand rubles	**≥ 413,2**	Projection 2 "Social Development"	I 2.1 Ratio of average per capita income of the population to the subsistence minimum, times	**≥ 3,5**	Projection 3 " Environmental Development"	I 3.1 Discharge of polluted wastewater, thousand m3 sq.km	**≤ 0,3**
	I 1.2 Annual inflation rate, %	**≤ 6**		I 2.2 The ratio of average pension to average wage, %	**≥ 40**			
	I 1.3 Investments in fixed assets, % of GRP	**≥ 25**		I 2.3 Unemployment level according to the ILO methodology, %	**≤ 4**		I 3.2 Emissions of pollutants into atmospheric air from stationary sources, t. sq. km	**≤ 0,5**
	I 1.4 Industrial enterprises fixed assets depreciation rate, %	**≤ 50**		I 2.4 Life expectancy at birth, years	**≥ 75**			
	I 1.5 Foreign trade balance, % of GRP	**≥ -4 до ≤ 8**		I 2.5 Size of housing per resident, sq/m	**≥ 25**		I 3.3 Reforestation (share of restored forests, %)	**≥ 0,15**
	I 1.6 Balance of the region's consolidated budget, % of GRP	**≥ -3 до ≤ 4**		I 2.6 Funds for healthcare, education and social support, % of GRP	**≥ 15**			

The analysis of the level of economic security of the Kaliningrad region made it possible to identify the main trends in socio-economic development, demonstrate the degree of SDGs achievement and identify threats to economic security.

During the analysis, a research algorithm was determined, representing a sequence of stages, including:

- Stage 1 – selection of an object and determination of the composition of indicators for assessing economic security;
- Stage 2 - collection of statistical data for calculating indicators and determining threshold values of indicators for the selected research object;
- Stage 3 - calculation of indicators and assessment of the level of economic security for each of the indicators when comparing their actual values with threshold values;
- Stage 4 - assessment by spheres of life activity and economic security in general for the selected object of study, determination of the prevailing trends in the socio-economic development of the region in terms of SDGs achieving, which is described in more detail in [11];

- Stage 5 – identification of threats to the economic security of the region that are relevant to the period of study and that impede the SDGs achievement;
- Stage 6 - development of strategic measures to ensure the neutralization of threats to the region's energy security and contribute to the SDGs achievement.

The obtained results are presented below in the context of three projections of the regional energy security system and the ESG concept (according to Fig. 2 presented above).

Table 3. Results of the analysis of the Kaliningrad region economic indicators in the projections "Economic development", "Social development" and "Environmental development"

Indicator	Year					Thresholds	Matching result (presence of threats)
	2018	2019	2020	2021	2022		
Projection "Economic Development"							
I 1.1	494,10	517,2	540,9	659,7	675,0	≥ 413,2 (t.roubles)	absence
I 1.2	4,8	2,8	5,1	4,2	3,1	≤ 6 (%)	absence
I 1.3	32,3	19,5	20,1	15,2	20,2	≥ 25 (% of GRP	presence
I 1.4	31,6	40,0	39,6	39,9	41,8	≤ 50 (%)	absence
I 1.5	-1,2	-1,2	-0,9	-0,9	-	≥ -4 till ≤ 8 (%of GRP)	absence
I 1.6	0,48	0,17	0,03	0,85	-0,02	≥ -3 till ≤ 4 (% of GRP)	absence
Projection "Social Development"							
I 2.1	2,53	2,55	2,50	2,65	2,38	≥ 3,5 (times)	presence
I 2.2	40,13	39,82	40,86	38,71	37,36	≥ 40 (%)	presence
I 2.3	4,7	4,4	5,9	4,6	3,1	≤ 4 (%)	presence
I 2.4	72,92	73,56	72,93	70,99	73,05	≥ 75 (years)	presence
I 2.5	28,2	28,9	31,5	31,4	28,2	≥ 25 (sq.m)	absence
I 2.6	7,56	7,89	9,31	7,48	8,06	≥ 15 (% of GRP)	presence
Projection "Environmental Development"							
I 3.1	0,49	0,53	0,52	0,45	0,44	≤ 0,3 (m3sq. km)	presence
I 3.2	1,58	1,51	1,37	1,52	1,79	≤ 0,5(t.sq.km)	presence
I 3.3	0,18	0,13	0,07	0,05	0,09	≥ 0,15 (%)	presence

Source: compiled by the authors using materials[12]

Further we will characterize the main emerging trends in the development of the region from the perspective of the concept of sustainable development in the assessed projections (Table 3).

4 Results and Discussion

Projection "Economic Development" - for most of the analyzed indicators, the calculated actual values do not go beyond the threshold, which indicates the absence of threats to regional economic security during the analyzed period of time. These indicators include:

- GRP per capita, which throughout the entire five-year period exceeds a given threshold value - at least 413.2 thousand rubles;
- annual inflation rate, which also does not exceed the threshold value of 6%, although it should be noted that it is closest to it in the region based on the results of 2020;
- the degree of industrial enterprises fixed assets depreciation, which also does not reach the critical level of 50%;
- foreign trade balance as a percentage of GRP, which is although negative, indicating an excess of imports over exports throughout the entire period under study, is still within the recommended threshold range. However, this ratio indicates the existence of the region's dependence on imports, which can be regarded as a threat to economic security;
- the actual value of the indicator "balance of the consolidated budget of the region in % of GRP" indicates the following: in 2018–2021 the regional budget was executed with a surplus, we saw a positive balance, but in 2022 the value was negative, there was a budget deficit, which also affected the economic development of the region;
- for the indicator "investment in fixed assets as a percentage of GRP", we note that only in 2018 did it exceed the recommended value of 25%, all other analyzed periods, 2019–2022, its values were much lower than the threshold, which indicates the existence of a threat to economic security for the economy of the Kaliningrad region.

Projection "Social Development" - Is especially relevant for the Kaliningrad region, because the values of five out of six ES indicators do not correspond to the recommended threshold values. Only in terms of housing provision, the population of the region, the actual value according to statistical reporting exceeds the recommended threshold.

Projection "Environmental Development" - Calculated and analyzed indicators in none of the areas do not meet the threshold values, which indicates the presence of threats to environmental safety in the region and requires the adoption of priority measures to neutralize them.

The trends identified through the analysis make it possible to see not only the presence of certain threats, but also the presence of "problem zones", "zones of challenges and risks", which can subsequently develop into real threats and in the future have a negative impact on the socio-economic situation in the region, threatening the achievement of territory sustainable development goals. Also, as a result of the analysis, it becomes possible to highlight areas for strengthening the economic security of the region, which plays an important role in ensuring a higher level of the hierarchical system - national economic security, taking into account SDGs defined in the Russian Federation.

Thus, we believe that the imported goods, including food, dependence problem is extremely relevant for the region. This negatively affects the ability to achieve SDGs number 2, 11 and 12, affecting both economic and social aspects. Considering the exclave position, sanctions, increased costs for the delivery of raw materials and supplies by sea and the inability to use alternative transport routes, complex logistics and other problems characteristic of the Kaliningrad region, complemented by low investment activity in terms of updating enterprises and industries fixed assets that form the basis of the region's economy, it is possible to predict the future growth of existing threats, as well as the development of potential risks, challenges and dangers.

Analysis of projections of social and environmental development not only proves the presence of threats to economic security in the region, but also increases concerns about achieving SDGs in the context of economic growth and social justice, also confirming existing threats to food security in the region, complementing the criterion of physical availability of food for of the population by a decrease in the level of its economic accessibility.

Concerns about achieving SDGs number 6, 14 and 15 cannot be ignored, which is due to the presence of problems and threats in terms of ensuring the environmental safety of the region.

As a result of the study, it was proven that it is possible to take into account the information and analytical tools of economic security, including a set of individual indicators, to assess the feasibility of goals within the framework of the concept of sustainable development when analyzing any of the ecosystem subsystems, detailing the level of the country, regions, individual organizations and individuals. This clarifies the theoretical and methodological approaches used in the analysis, thereby increasing the validity and quality of decision making.

References

1. United Nations Environment Programme International Resource Panel: Policy Coherence of the Sustainable Development Goals: A Natural Resource Perspective. https://wedocs.unep.org/20.500.11822/9720. Accessed 17 Nov 2023
2. Environmental Protection Agency: Environmental Investments: the Cost of a Clean Environment: Report of the Administrator, Island Press, Washington, Covelo, California (1993)
3. Panayotou, T.: Green Markets: The Economics of Sustainable Development. ICS Press Institute for Contemporary Studies, San Francisco, USA (1993)
4. Mnatsakanyan, A., Pobegaylo, M.: Development of fisheries in the Kaliningrad region in the aspect of ensuring food security. Baltic Econ. J. **3**(39), 31–42 (2022)
5. Evenson, R.E.: Research and Extension in Agricultural Development: Scientific Edition, ICS Press, San Francisco (1992)
6. Ilyina, O.: State documents of strategic planning as a source of information for analyzing threats to the economic security of an enterprise when carrying out foreign economic activity. Baltic Econ. J. **3**(43), 26–36 (2023)
7. Order of the Government of the Russian Federation dated July 14, 2021 N 1912-r as amended on December 30, 2023. In: On approval of the goals and main directions of sustainable (including green) development of the Russian Federation (2023)
8. Committee on World Food Security: Report of the 41st Session of the Committee on World Food Security. Rome, 13–18 October 2014. https://www.fao.org/3/mm369e/mm369e.pdf. Accessed 15 Jan 2024
9. FAO: The State of World Fisheries and Aquaculture 2022. In: Towards Blue Transformation, Rome, FAO (2022)

10. Food and Agriculture Organization of the United Nations: The State of World Fisheries and Aquaculture, 2016: Contributing to Food Security and Nutrition for All. Technical Report, Rome. https://www.fao.org/3/i5555e/i5555e.pdf. Accessed 15 Jan 2024
11. Mnatsakanyan, A., Kuzin, V.: Trends and features of socio-economic development of the Kaliningrad region. In: International Scientific Conference Competitive, Sustainable and Secure Development of the Regional Economy, Response to Global Challenges (2018)
12. Kaliningrad Region in Numbers. 2023: Brief statistical collection/Kaliningradstat, Kaliningrad (2023)

Financial Model of Industrial Cluster: Features of Value Chain Creation and Development

Ruslan Polyakov(✉) and Vladimir Kuzin

Kalingrad State Technical University, Kaliningrad, Russia
ruslan.polyakov@klgtu.ru

Abstract. This paper shows the evolution of approaches to the study, formation and development of clusters. The structural question became how the cluster caused the transition from post-war recovery to economic growth and accelerated development. One of the central issues was the consideration of financial aspects of cluster organization and operation. This issue is studied from two different points of view: firstly, on the basis of consideration of territorial cluster and, secondly, industrial cluster. The paper formulates the definition of the concept of cluster finance, classifies the financial relations of the cluster and investigates the possibilities of financial support for the formation and development of the cluster. Financial models of territorial and industrial clusters are analyzed. Understanding of the financial model of cluster functioning and its peculiarities allows to optimize the development of both the cluster itself and the accompanying infrastructure.

Keywords: Cluster · Territorial Cluster · Industrial Cluster · Cluster Finance · Financial Model · Financial Support Efficiency Indicators

1 Introduction

In domestic scientific practice the concept of cluster appeared in the 70s of the last century. It was introduced into scientific circulation by specialists in economic geography Gorkin A.P., Smirnyagin L.V.. [1], who considered the conditions and factors of production location. The cluster approach became the development of methods practiced within the framework of the Council for the Study of Productive Forces (COPS). The result of the SPSS activities were decisions on the location of large production facilities. It should be noted that the placement of production facilities was most often based on military security requirements and could not take into account market requirements under the planned economy, since market conditions were distorted at that time.

Due to the specificity of scientific interests in the conditions of planned economy, many economic aspects remained outside the scope of consideration. Such aspects include finance, which in the framework of economic theories of socialism, played an auxiliary role. It is enough to remember that the purchasing power of money in this period was not independent and had to be confirmed by funds and limits allocated in the planned order. Thus, money did not fully fulfill the function of the measure of value.

R. Polyakov (Ed.): EcoSyst 2023, LNNS 1092, pp. 121–128, 2024.
https://doi.org/10.1007/978-3-031-67354-2_14

It is also necessary to mention an earlier theoretical approach based on territorial-production complexes (TPK). According to the Big Russian Encyclopedia, TPC is understood as "a form of territorial organization of production, one of the main concepts of Soviet economic geography and regional economics". The concept of TIC was introduced into scientific circulation by N. N. Kolosovsky, a specialist in economic geography in the 1940s. According to him, TIC is "an economic combination of enterprises in one industrial point or in the whole area, which achieves a certain economic effect due to the successful selection of enterprises in accordance with the natural and economic conditions of the area, with its transportation and economic-geographical position" [2].

The advantage of the TPC approach is undoubtedly the appearance of economic effect as one of the target functions. At the same time, the TPC approach is based on the national economic efficiency, This concept in its development is reflected in the Methodological Recommendations for assessing the effectiveness of investment projects and their selection for financing, which states that "indicators of national economic efficiency reflect the effectiveness of the project in terms of the interests of the national economy as a whole, as well as for the participating regions (subjects of the federation), industries, organizations and enterprises involved in the implementation of the project" [3]. Thus, the national economic efficiency allows taking into account the market requirements. However, the formation of the TIC concept be-came the implementation of the planned economy model in its territorial dimension does not take into account the aspects of competition and globalization of economic activity.

In the 70s, Swedish economists K Fredriksson and L. Lindmark began to use the term "cluster" to designate clusters of enterprises in space. This happened within the framework of research on the problems of regional economy". [4]. The authors, based on the collected data, studied, based on the ideas of the importance of specialization and the proximity factor, the structure and scale of subcontracting systems in the region. The role of both large and small enterprises was considered.

Interest in structural issues has increased since the late 1960s. This interest was largely due to the transition from post-war recovery to economic growth and accelerated development. Subsequently, research has focused on the location and internationalization of small firms' business [5].

2 Materials and Methods

The above-mentioned studies were theoretical in nature and the concept of cluster was not much in demand. In addition, the transition from a planned economic sys-tem to a market one. This required taking into account market factors in theoretical research and revising the role of finance.

In this situation, M. Porter's application of the concept of cluster to describe the factors of competitiveness of groups of enterprises. In his opinion "today's economic map of the world is dominated by what I call clusters: critical masses (in one place) of unusual competitive success in certain areas. Clusters are a striking feature of virtually every national, regional, state, and even metropolitan economy, especially in the more economically developed countries. But clusters are not unique; they are highly typical - and therein lies the paradox: sustainable competitive advantage in the global economy is increasingly based on local things - knowledge, attitudes, motivation - that distant rivals cannot match" [6].

According to M. Porter, clusters are geographical concentrations of interrelated companies and institutions in a certain area. Clusters encompass many related industries and other organizations important for competition. These include, for example, suppliers of specialized resources such as components, equipment and services, as well as suppliers of specialized infrastructure.

On the one hand, in modern economic literature, a cluster is understood as geographical concentrations of interrelated companies and institutions in a certain type of economic activity in a certain area (or closely related types of such activities). Thus, a cluster is primarily a territorial formation. On the other hand, within the framework of the federal law "On Industrial Policy in the Russian Federation" in 2014 the concept of an industrial cluster was introduced as a set of subjects of activity in the sphere of industry, connected by relations in the specified sphere due to territorial proximity and functional dependence and located in the territory of one subject of the Russian Federation or in the territories of several subjects of the Russian Federation.

Such a definition, in our opinion, from the spatial point of view is rather blurred, since the location of the cluster on the territories of several subjects of the federation creates problems of financial incentives at the territorial level. However, in order to adequately consider this problem, it is necessary to define the concept of cluster finances.

Currently, finance is understood as a set of economic relations arising in the process of formation, distribution and use of centralized and decentralized funds of money. This definition is not the only one, but we have chosen it because it contains economic relations, i.e. relations between people on the issue of using available re-sources. Financial resources are understood as monetary funds with different terms of receipt.

When forming the definition, it is possible to take into account the peculiarities of the subject of research. As an example, we can consider the definition of sectoral finance. Here it is necessary to recall the definition of the industry, which was formed in the framework of the use of OKONKh - the all-union classifier of industries of national economy, which was in effect until January 01, 2005. Within the framework of OKONKh the industry was understood as a set of similar types of entrepreneurial activity having common features. At present, OKVED - All-Russian Classifier of Economic Activities - is in force, but the concept of industry is still used in scientific practice.

Sectoral finance is that part of finance, which is located in the sphere of production of economic goods and may well exist outside the state. [7]. Such a definition is not operational, that is, it does not give the possibility of its use in practical activity. That is, it is not suitable for solving the problems of our research.

3 Discussion

Within the framework of forming the definition of cluster finances, let us consider financial relations related to the cluster. Financial relations, as applied to the cluster, are economic relations between the entities that are members of the cluster and inter-act with them, which are associated with the formation, distribution and use of funds to meet the needs of the cluster (organizations and individuals who are members of the cluster and interact with them). The peculiarities of cluster finances are that they include both decentralized and centralized finances. Classification of financial relations of the cluster is presented in Fig. 1.

The figure shows that the financial relations of the cluster are divided into two groups. The first group - centralized finances, which include the relations related to the calculation and payment of taxes by the organizations included in the cluster and interacting with the cluster enterprises, as well as the relations caused by budgetary support of the cluster members.

The second group includes relations related to decentralized finance. This group includes relations caused by the purchase of raw materials and components from organizations outside the cluster, as well as the sale of goods and services to organizations external to the cluster. This group also includes relations related to the payment of salaries. A separate group of financial relations includes various transactions be-tween organizations that are members of the cluster.

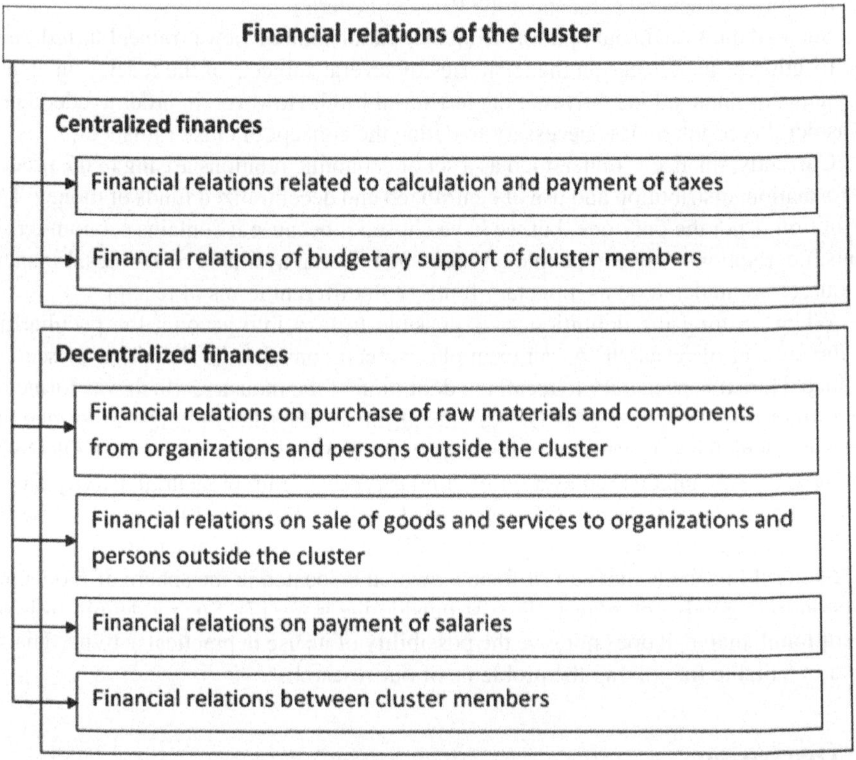

Fig. 1. Financial relations of the cluster.

In order to be able to forecast the results of the cluster's activities and assess the financial results within a given period, it is advisable to build its financial model. For the purposes of our work, the financial model will be understood as a scheme reflecting the economic state of the cluster and its member organizations. The model may include financial indicators characterizing the efficiency of the use of financial re-sources allocated to support the cluster.

In order to explain the main content of the model, it is necessary to explain the peculiarities of the organization of the financial system of the Russian Federation, in terms of centralized finance, in relation to the topic of the article. We proceed from the fact that the spatial element for our consideration is the subject of federation (region), which is a state entity. Financial powers of the subject of federation extend to its territory and persons who are its residents. Thus, there are no legal grounds to finance, including providing financial support to economic entities not related to the relevant subject of the federation. Further we will see that federal financial support is possible for industrial clusters whose member enterprises are located on the territory of several subjects of the federation. However, from our point of view, the main motivating factor for the formation and support of clusters at the regional level is the prospect of financial development of the territory.

Such development is ensured through the involvement of additional enterprises in the cluster, whose production volume increases due to their inclusion in the cluster value chain, for example, by producing elements, components and assemblies that were previously produced outside the region. In addition, the cluster is a system whose production capacity is greater than the sum of the production capacities of the set of enterprises that make up the cluster.

This is achieved through specialization of individual enterprises within the cluster and economies of scale. In this case, the cluster can be presented as a large economic entity consisting of a large number of small and small enterprises. The financial structure of the cluster, taking into account the interest of public authorities in its development, is close to public-private partnership projects [8].

On the other hand, the fractional structure of the cluster allows more flexible adaptation to the negative effects of the external environment. Thus, in case of unfavorable changes in the market conditions, leading to a reduction in demand for the cluster products, its structure consisting of a significant number of small and medium-sized enterprises allows to preserve the production and commercial potential of enterprises. Small and medium-sized enterprises, being more flexible than large ones, are quicker than large ones to switch to new types of products and switch to new markets. At the same time, the cluster, as a whole, following the individual enterprises included in it, gradually rebuilds itself, remaining a stable system.

4 Research Results

The systemic approach to the cluster remains beyond our consideration, but it should be noted that the cluster is based on the key property of the system - synergy, which is the fact that the result of the cluster as a set of enterprises is greater than the sum of the results of the enterprises' activities if they did not make up the cluster. From the point of view of finance, it means that within the cluster the financial performance of the enterprises is higher compared to that shown by the enterprise when working individually. From the point of view of regional authorities, this allows to ensure financial sustainability of the enterprise, which is an employer and tax payer. This explains the necessity and expediency of support of clusters by regional authorities.

The financial structure of the cluster has peculiarities [9], which is reflected in the financial model of the regional cluster shown in Fig. 2. It should be noted that the authorities of the subject of the federation have other objectives, including social ones. For example, to ensure employment close to full employment. This leads to the fact that they support the creation of enterprises that create jobs. In this case we focused on the financial aspects of the cluster.

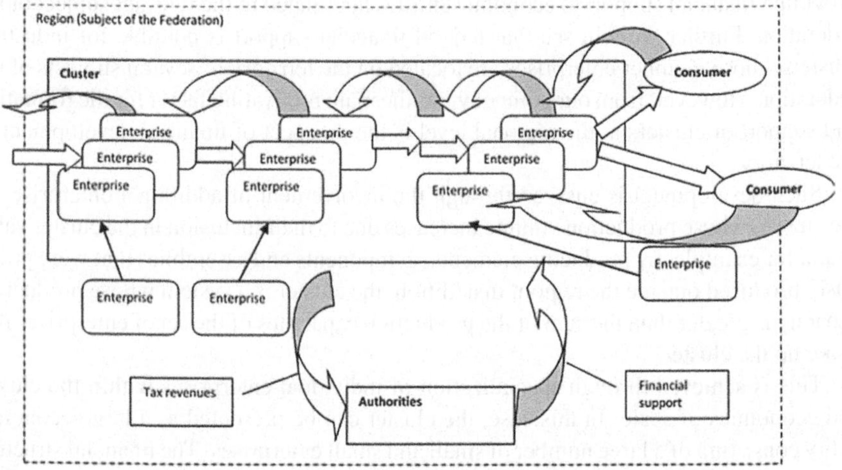

Fig. 2. Financial model of the territorial cluster.

In the financial model of the territorial cluster, one of the important mechanisms of development is the involvement of additional enterprises in its composition, developing value chains in the territory. This increases the competitiveness of the territory and concentrates financial flows within the region. Thus, clusters and cluster policy acts as a driver of regional development [10].

This allows the authorities with a relatively small level of budgetary support to form a sustainable community of economic entities and through this to ensure a steady flow of tax revenues. The model reflects only direct financial flows. Research results show that individual, including financial, indicators with positive dynamics are positively reflected in various aspects of life in the region [11]. The development of modern digital methods of accounting and tracking of financial transactions allows us to monitor the achievement of financial goals of the authorities and assess the effectiveness of their achievement [12, 13].

The finances of territorial cluster and industrial clusters in the Russian Federation have different models of functioning. Although a territorial cluster can have features characteristic of an industrial cluster. The cluster of shipbuilding and ship repair in the Kaliningrad region can serve as an example of such a cluster [14]. For a territorial cluster, the financial model is limited to the territory of the subject of the federation and, as it is shown in Fig. 2, and includes enterprises, with few exceptions, enterprises registered and operating in the territory of the region. In turn, the industrial cluster, as it is shown in Fig. 3, actually includes enterprises from different subjects of the federation.

Unlike a territorial cluster, an industrial cluster is created in order to ensure the production of necessary products. That is, if there are financial goals, they are not the main ones when creating an industrial cluster. At the same time, financial goals for enterprises are invariable and arise from its essence. An enterprise is an organization created for the purpose of profit making in the interests of the founders. This forms a dualism of enterprise goals within the industrial cluster. The enterprise has financial goals, and the cluster has the creation of certain products.

This determines the peculiarities of the financial model of the industrial cluster. It can include enterprises from different subjects of the federation. The center of profit formation (or other taxable base) in the formation of products can be formed based on the interests of persons who are not included in the number of cluster participants or authorities. This reduces the interest of regional authorities to support the industrial cluster. At the same time, the main stakeholder remains - the federal authorities, which is reflected in the financial model of the industrial cluster shown in Fig. 3.

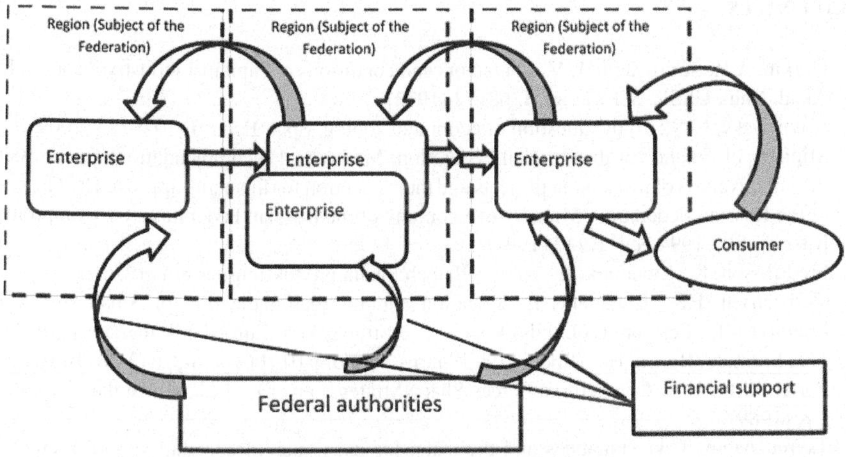

Fig. 3. Financial model of the industrial cluster.

The peculiarities of the financial model of the industrial cluster are due to the fact that in order to ensure the production of the required products it is necessary to ensure the achievement of financial goals of the enterprises included in the cluster. For example, the required rate of profit. If this is impossible in the current market conditions, financial stimulation of enterprises is required. This can occur at the stage of cluster creation, namely the formation of stable economic relations between enterprises, the development of infrastructure necessary for the production of required products, as well as the implementation of projects and programs of the cluster.

The model in Fig. 3 does not include tax revenues due to the fact that the main goal of industrial cluster formation is the creation of necessary products, and the increase in tax revenues is not fundamental for this type of clusters, although the successful achievement of the goal is achieved and the growth of tax revenues.

5 Conclusion

From the point of view of finances, a special case of a cluster is optimal, namely a production cluster created within one territory. In this case, the interests of regional and federal authorities coincide. For enterprises, this creates an opportunity to receive financial support to stimulate the production of required products from the federal authorities and support aimed at ensuring the sustainability of the regional industry.

Understanding of the financial model of cluster functioning and peculiarities of functioning of the financial system of territorial and industrial cluster allows to justify managerial decisions on financial (budgetary) support of clusters during their creation, infrastructure development and implementation of projects and programs with-in the cluster.

References

1. Gorkin, A.P., Smirnyagin, L.V.: On factors and conditions of capitalist industry location. Izv. Akad. Nauk USSR. Ser. Geogr. **1**, 68–72 (1973)
2. Kolosovsky, N.N.: To the question of economic zoning. Spat. Econ. **1**, 102–123 (2009)
3. Ministry of Finance of the Russian Federation: Methodical recommendations for assessing the effectiveness of investment projects and their selection for financing approved by Gosstroy of the Russian Federation, Ministry of Economy of the Russian Federation, Goscomprom of Russia 31.03.1994 N 7–12/47 (1994)
4. Fredriksson, K., Lindmark, L.: Nationella och lokala produktionssystem (1976)
5. Christensen, P.R., Lindmark, L.: Location and Internationalization of Small Firms. In: Lundqvist, L., Persson, L.O. (eds) Visions and Strategies in European Integration, pp. 131–151. Springer, Berlin, Heidelberg (1993). https://doi.org/10.1007/978-3-642-78178-0_9
6. Porter, M.E.: How Competitive Forces Shape Strategy, pp. 133–143. Macmillan Education, UK (1989)
7. Tsyrenzhapov, C.D.: Finances and the state: topical issues of correlation and interaction. Siberian Legal Bull. **1**(88), 38–43 (2020)
8. Karlov, A.M., Mnatsakanyan, R.A.: Method of determining parameters of government participation in financing PPP projects in the sphere of RHC. Baltic Econ. J., 340–346 (2022)
9. Merzlov, I.Y.: Financial model of construction and functioning of a cluster. Vopr. Ekon. **5**, 17–25 (2007)
10. Polyakov, R.K., Durov, A.V.: The role of clusters and cluster policy in the development of the Kaliningrad region. Reg. Econ. Manag. Electron. Sci. J. **2**(50), 19 (2017)
11. Kohan, A.N., Kuzin, V.I., Karlov, A.M.: Interrelation of sustainable development and food security indicators at the regional level. Baltic Econ. J., 18–34 (2023)
12. Sergeev, L.I., Sergeev, D.L.: Interbudgetary forms and digital methods of regulation of their relations. Baltic Econ. J., 87–107 (2022)
13. Sabinina, A.L., Kuznetsova, N.N.: Ways of assessing the effectiveness of the use of different sources of financing. Izv. Tula State Univ. Econ. Legal Sci. **4–1**, 3–11 (2015)
14. Polyakov, R.K., Kuzin, V.I.: Cluster approach in the development of shipbuilding and ship repair in the Kaliningrad region. Baltic Econ. J., 132–147 (2023)

Financial Approach to Strategy for Environmentally-Socio-economically Oriented Fisheries Development

Albert Mnatsakanyan(✉) and Alexander Kharin

Kaliningrad State Technical University, Kaliningrad, Russian Federation
mag@klgtu.ru

Abstract. Developing a fisheries strategy based on the principles of sustainable development is a complex task that has not yet received an adequate solution. The main problem is the promiscuity of sustainability criteria, aimed at achieving economic, social and environmental goals simultaneously. The article gives a financial approach to fisheries optimization. Its advantage is the possibility of unifying optimality criteria. Based on a generalization of research results in the field of fisheries economics and the well-known provisions of neoclassical economic theory, a theoretical model is proposed. The model simplifies the solution of fisheries management problem from the standpoint of sustainability. The advantages of the model are the use of a limited set of parameters covering the main ecosystem, social and economic aspects of fisheries, as well as the ease of practical implementation and results monitoring. The article provides recommendations regarding some special cases of implementing a sustainable fisheries strategy, directly determined by the content of the model, and also indicates the limitations of the model and possible ways of their elimination.

Keywords: Sustainable Development · Fisheries · Strategy · Modeling

1 Introduction

The problem of the fisheries transition to a model of sustainable development, environmentally, socially and economically balanced is quite widely discussed in modern literature. Recently, this problem has also been the focus of attention of many governments and international organizations. Such a great interest in finding new ways to organize fisheries occurred due to the inability of traditional management methods to cover the full range of consequences caused by fisheries management. On the one hand, the limited ability of aquatic ecosystems self-reproduction requires solutions to prevent their depletion and degradation in the interests of current and future generations. On the other hand, fish and seafood are an important food resource that has a significant impact on people's lives, thus "the sustainability of fisheries and aquaculture is a fundamental condition for food security and nutrition" [1, 2]. In the modern understanding, sustainable fishing is a way of conducting fishing activities in which a compromise cab be

R. Polyakov (Ed.): EcoSyst 2023, LNNS 1092, pp. 129–135, 2024.
https://doi.org/10.1007/978-3-031-67354-2_15

achieved between social and economic interests on one side and the tasks of potential of exploited resources saving at the population and ecosystem levels on the other. These key features are reflected in the UN-supported Sustainable Development Goals [3], which, among other things, define the role of fisheries in this process and are intended to serve as the basis for the corresponding strategies of countries, international organizations and economic entities.

2 Materials and Methods

Different authors and organizations propose different ways to identify factors that impede sustainability and select appropriate measures to compensate for these threats. As a rule, they are all based on the idea that sustainable fishing is impossible without comprehensive consideration of all its results and consequences of an environmental, economic and social nature [4]. In particular, the ESG concept has recently become increasingly popular in sustainable development problem solving at the corporate level. This management technology is based on the principle when companies strive for consistency in the goals of corporate development, rational use of natural resources and achievement of social results [5]. However, no matter the methods for optimal management decisions, they all inevitably face a common problem - defining and ranking sustainability criteria based on environmental and socio-economic development goals. Moreover, these criteria can be alternative and even opposite in nature. In addition, decision makers also inevitably face the problem of intertemporal choice; they need to answer the question: which benefits are more preferable – today's or future ones.

Typically, the theory of economic well-being state as the methodological basis for choosing the optimal option among many possible strategies in the condition of limited resources. In the general case, according to this theory, optimality means that production corresponds to one of the points on the boundary of a social possibilities set, at which maximum economic and social effects are achieved, thus, the greatest social utility is ensured [6]. In the case of fishing, this state can be interpreted as maximizing net social benefit, defined as the sum of the total benefits obtained from fishing activities minus all costs associated with this activity, including explicit and hidden environmental and social costs from the exploitation of a fishery resource.

Although the above statement quite accurately reflects the essence of the fisheries optimization problem, its practical solution is associated with serious difficulties. The main problem is the above-mentioned difference in the goals of stakeholders and, accordingly, the heterogeneity of optimality criteria. One of the well-known ways to eliminate this heterogeneity is the financial approach. The essence of this approach is to reduce all effects from the use of resources to universal cost indicators. In a market economy, the basis of metrics for calculating such indicators is usually prices, information about which is available and provides fairly wide opportunities for analysis. However, all market prices must be complete, which means each product price participating in production chains must be determined in order for this method to give an adequate result. In addition, the prices must correctly reflect the social value of each unit of the product. In a developed and normally functioning market economy both of these conditions are largely satisfied. But in cases where market relations are not sufficiently developed or they are severely distorted, adjustments to apparent prices are necessary to be made. This modifying prices procedure in order to use in optimality calculations is currently standard; in particular, it has been worked out in quite detail theoretically and is widely used in cost-benefit analysis (see, for example, [7]). In a market economy, there are also frequent situations when researchers are forced to turn to constructing artificial analogues of market prices. Typical goods that require price design are natural resources, as well as other assets, in respect of which, for various reasons, private property rights cannot be established [8]. This circumstance determines the absence of appropriate markets and, consequently, clearly defined market prices for these goods. In this case, the equivalent of the true price is the so-called shadow price.

Fish resources are precisely such a product that does not have a clearly defined price. The amount of fish resources, in relation to the fishing economics, is usually designated by the term "biomass". It should be noted that, along with price, the volume of biomass is considered as one of the key parameters when choosing a sustainable fisheries model, since no matter how effectively fisheries are managed in terms of economic and social results, they inevitably negatively affect aquatic ecosystems. This non-cost parameter, which in general characterizes the ecological state of a fishery resource, occupies one of the central places in most financial metrics describing sustainable fisheries.

The shadow value of biomass is, in essence, the price that the only user of the resource is willing to pay for increasing biomass by one more unit. In the case of an optimal fishery, it is equivalent to the net benefit that would accrue to all users if the sustainable biomass increased by one unit. This value represents the monetized value of the fishery resource, assessed from the point of view of society. If the volume of biomass is small and the state of this resource is moving toward depletion, its shadow value increases. In contrast, when fish stocks are abundant, the shadow value of biomass is relatively low. Thus, in a fundamental sense, the shadow value of a fishery resource is a measure of its relative rarity [9]. This property allows us to understand the nature of optimal and sustainable fishing and the motives of behavior of all stakeholders influencing strategic decision-making.

3 Results

The conclusion made above serves as the basis for interpreting the shadow value of a fishery resource as social costs caused by the extraction of one unit of its biomass. These costs reduce the total benefit from the fishery. Thus, the net social price of a unit of extracted resource can be defined as its market price minus its shadow value:

$$s = p - \lambda,$$

where s – true social price of a resource unit, p – market value, λ – biomass shadow value.

In traditional fisheries models, the shadow value of biomass is an imaginary value; it is not paid to anyone, and fishing companies actually receive the market price for the fish caught and sold. However, in the case of a sustainable fishery, which is optimal from a social point of view, the situation appears as if there was a true social price for fish. Following this logic, the problem of optimizing fishing activity can be reduced to the classical economic problem of maximizing profit by controlling only two parameters - fishing effort e and biomass size q:

$$\max_{e,q}(p \cdot Y(e, q) - C(e)),$$

provided that

$$G(q) - Y(e, q) = 0, \tag{1}$$

where $G(q)$ – natural increase in fish resource biomass, $Y(e,q)$ – volume of fishery resource withdrawal (production), $C(e)$ – production costs.

A possible way to solve this problem is to state the Lagrange equation of the form:

$$L(e, q, \lambda) = p \cdot Y(e, q) - C(e) + \lambda \cdot (G(q) - Ye, q)),$$

where λ – Lagrange multiplier, which, with an optimal solution, takes on the meaning of the shadow value of biomass.

In addition, the necessary conditions for solving problem (1) are:

$$p \cdot Y_e - C_e - \lambda \cdot Y_e = 0, \tag{2}$$

$$p \cdot Y_q + \lambda \cdot (G_q - Y_q) = 0. \tag{3}$$

Equation (2) can be interpreted as a basic condition for sustainable fisheries - a condition in which marginal benefits, adjusted for the shadow value of biomass, must correspond to marginal costs:

$$(p - \lambda) \cdot Y_e = s \cdot Y_e = C_e.$$

Excluded λ from Eqs. (2) and Eq. (3) and performing a simple rearrangement, we obtain the basic optimality conditions:

$$G_q + Y_q \cdot C_e/(p \cdot Y_e - C_e) = 0,$$

$$G(q) - Y(e, q) = 0. \tag{4}$$

Solving Eq. (4) for e and q allows us to determine the target values of two key parameters that define the vector of sustainable fisheries policy. And Eq. (3) gives the answer to the value of the shadow value of biomass:

$$\lambda = p \cdot Y_q / (Y_q - G_q).$$

Note that this value tends to infinity when $(Y_q - G_q)$ tends to zero. This is the situation that is typical for most modern fisheries that exploit fish resources whose biomass level is lower than that corresponding to the maximum sustainable yield (MSY) [10].

The resulting equations describe the conditions of static equilibrium of fisheries. However, the concept of sustainability does not make sense without taking into account the temporal context. Therefore, the fishery optimization model must include long-term sustainability conditions. As before, we assume that the goal of sustainable fisheries is to maximize the magnitude of net benefits. However, now this goal implies solving a dynamic optimization problem designed to select a trajectory of fishing effort that maximizes the current value of the net benefit from fishing over an infinitely long management time horizon:

$$\max_{\{e\}} \int_0^{\infty} (p \cdot Y(e, q) - C(e)) \cdot e^{-rt} dt,$$

provided that

$$\dot{q} = G(q) - Y(e, q). \tag{5}$$

One way to solve this problem is to determine the Hamilton function and subsequent derivation of the necessary conditions (see, for example, [11, 12]). In the case we are considering, the Hamiltonian is the sum of the aggregated current value of the net profit of fishing companies and the shadow value of biomass in a steady state, taking into account bioeconomic restrictions (natural growth of biomass and the volume of its total extraction):

$$H(e, q, \mu) = p \cdot Y(e, q) - C(e) + \mu \times (G(q) - Y(e, q)),$$

where μ – dynamic Lagrange multiplier.

From the Hamilton function we can obtain the conditions for solving the problem (5):

$$p \cdot Y_e - C_e - \mu \cdot Y_e = 0,$$

$$\dot{\mu} - r \cdot \mu = -p \cdot Y_q - \mu \cdot (G_q - Y_q),$$

$$\dot{q} = G(q) - Y(e, q).$$

Having imposed equilibrium conditions $\dot{q} = \dot{\mu} = 0$ and excluded the dynamic Lagrange multiplier μ from the first two equations, we obtain the main optimality conditions in equilibrium:

$$G_q + Y_q \cdot C_e / (p \cdot Y_e - C_e) = r,$$

$$G(q) - Y(e, q) = 0. \tag{6}$$

Solving these equations for e and q allows us to determine the values of the two main parameters of the fisheries policy that correspond to the conditions of dynamic equilibrium and are most consistent with the principles of sustainability.

In practice the nature and form of optimal solutions can vary greatly depending on the characteristics of each resource and the specifics of the fishery organization. Nevertheless, the results obtained allow us to draw several general conclusions about possible fishing strategies. Thus, if the biomass of a resource is below the optimal sustainable level, then the optimal dynamic policy requires initially relatively low fishing effort (and, therefore, catches) in order to ensure restoration of biomass and, conversely, relatively high fishing effort if biomass is above the optimal sustainable level. It is also important to understand the main consequences that arise due to the difference between dynamic and static equilibria. The consequence of taking into account the discount rate in a dynamic model of optimal fishing is, as a rule, an increase in fishing effort and a decrease in the optimal equilibrium biomass compared to the static model. This leads to the fact that, at a sufficiently high rate, the optimal dynamic equilibrium can be located to the right of the point of maximum sustainable profitability based on the conditions of static equilibrium.

4 Results and Discussion

The presented approach represents a theoretical interpretation of the fisheries management problem based on the principles of the sustainable development concept. Some directions for solving this problem, based on the provisions of neoclassical economic theory, proposed in the works of scientists of the late twentieth century, can also be applied to manage sustainable fisheries. However, there are a number of important features that require consideration when developing a strategy for sustainable fishing in environmental, social and economic terms. First, if the biomass of a resource is below the minimum viable level, then it is clear that the only reasonable sustainable policy is to restore that resource. Second, if biomass is initially at a low but sustainable level, exploitation of the resource may be optimal. The criterion for making a decision in this case is the value of the discount rate (the higher it is, the more appropriate the fishery looks). However, in practice this situation is unlikely because, as noted above, most fisheries resources are exploited at levels well above MSY. In addition, as biomass decreases, its shadow value increases, which offsets the high discount rate and reduces the social benefit from the fishery. Finally, some fish stocks are preferable to be exploited on a periodic basis, especially those characterized by low rates of biological growth and high economic costs of organizing fishing. In this case, the optimal policy involves active fishing activities for a certain period of time, followed by a pause for restoration (natural or artificial reproduction) of the resource. Despite the fact that this mode of exploitation does not fully comply with sustainable development concept rules, such a fisheries policy can be considered sustainable in the ecosystem sense with some reservations.

It should also be noted that although the proposed approach is based on the ideas of ecosystem sustainability and maximizing fisheries contribution to the set of social opportunities, which is theoretically a prerequisite for improving the well-being of each member of society, it does not guarantee this at all. According to welfare theory, choosing the optimal fishing model is more of a social and political problem than an economic or environmental problem. It is clear that fisheries policy cannot ignore the wide range of issues associated with the distribution of benefits and costs arising from both the exploitation and the very existence of fisheries resources. Many of these benefits and costs are non-monetizable and cannot be reduced to a narrow financial indicator of the shadow cost of biomass. In cases where there is no reason to expect that the costs and benefits arising in the process of fishing activities will be distributed fairly, in accordance with the preferences prevailing in society, the proposed approach cannot give a satisfactory result. In this situation, optimal fisheries policy should be based on different, less formal approaches that better take into account social orientations and political attitudes. In particular, a possible alternative to optimization methods can be system dynamics models. The advantage of these models is the ability to describe the complex ecosystems behavior at a high level of abstraction, best reflecting modern ideas about sustainable fishing and, more than the financial approach, corresponding to the usual logic management of such systems [13]. In this case, the proposed approach can become one of the elements of a complex model that describes the behavior of financial and quasi-financial indicators. The presence of this element will eliminate some of the shortcomings of system dynamics [14], inevitable in the case of economic use of such complex objects as aquatic ecosystems; it will also give the process of these objects managing a more pragmatic nature and, thus, improve the quality of decisions made.

References

1. Committee on World Food Security: Report of the 41st Session of the Committee on World Food Security. Rome, 13–18 October 2014. https://www.fao.org/3/mm369e/mm369e.pdf. Accessed 15 Jan 2024
2. Food and Agriculture Organization of the United Nations: The State of World Fisheries and Aquaculture, 2016: Contributing to Food Security and Nutrition for All. Technical Report. Rome (2016). https://www.fao.org/3/i5555e/i5555e.pdf. Accessed 15 Jan 2024
3. Sustainable Development Goals. https://www.un.org/sustainabledevelopment/. Accessed 16 Jan 2024
4. Kolonchin, K.V.: On the way to sustainable development of marine industrial fisheries. Fish. Issues **23**(1), 5–15 (2022)
5. Tomkovich, A.V.: Using ESG principles in the development of the Russian fisheries industry. Baltic Econ. J. **1**(37), 56–64 (2022)
6. Feldman, A.M.: Welfare Economics, pp. 1–14. The New Palgrave Dictionary of Economics. London, Palgrave Macmillan UK (2017)
7. Layard, P.R.G., Glaister, S.: Cost-Benefit Analysis, 2nd edn. Cambridge University Press, Cambridge (1994)
8. Howarth, R.B., Farber, S.: Accounting for the value of ecosystem services. Ecol. Econ. **41**(3), 421–429 (2002)
9. Swartz, W., Sumaila, R., Watson, R.: Global ex-vessel fish price database revisited: a new approach for estimating 'missing' prices. Environ. Resour. Econ. **56**(4), 467–480 (2013)
10. FAO: The State of World Fisheries and Aquaculture 2022. Towards Blue Transformation. Rome (2022)
11. Conrad, J.M., Clark, C.W.: Natural Resource Economics: Notes and Problems, Cambridge University Press (1987)
12. Clark, C.W., Munro, G.: The economics of fishing and modern capital theory: a simplified approach. J. Environ. Econ. Manag. **2**, 92–106 (1975)
13. Mnatsakanyan, A.G., Karlov, A.M., Kharin, A.G.: Multi-criteria approach to decision making in the field of food security. Baltic Econ. J. **4**(36), 5–14 (2021)
14. Faster Capital: System Dynamics: Understanding Complex Systems with System Dynamics Model Simulation. https://fastercapital.com/content/System-Dynamics--Understanding-Complex-Systems-with-System-Dynamics-Model-Simulation.html. Accessed 18 Jan 2024

On Economic Indicators of the Global Financial Market

Olga Kuzmina[1(✉)], Maria Konovalova[1], and Tatyana Stepanova[2]

[1] Department of Economic Theory, Samara State University of Economics, Samara, Russia
pisakina83@yandex.ru

[2] Department of Economic Security, Kaliningrad State Technical University, Kaliningrad, Russia

Abstract. There is a huge variety of stock indices in the securities market, which are used both to assess the state of a particular industry and the state of the economy as a whole. Thanks to indices, it is possible to analyze the historical development of the stock market, to identify current trends, as well as to outline possible future paths of stock market development. Most economists consider stock indices to be important generalizing indicators of statistics. According to the indices it is possible to assess the state of the state economy as a whole, individual industries, make a forecast of them. Investors can use stock indices as an ideal model of diversification of their portfolio or as a tool for assessing the effectiveness of its management. The purpose of this study is to identify internal and external factors that affect the sensitivity of the index. The article provides a comparative analysis of Moscow Exchange indices, studies the historical dynamics of Russian and American indices, and identifies the factors that determine the degree of index sensitivity to certain economic changes. Of particular interest is the authors' justification of the differences in the degree of reaction of Russian and American indices to the dynamics of the same endogenous and exogenous factors.

Keywords: Financial Market · Stock Indices · Degree Of Sensitivity To Changes

1 Introduction

Stock indices are one of the tools used to analyze the dynamics of development of individual segments of the stock market and the stock market as a whole. Stock market analysts and investors use stock indices to assess the state of the stock market, to forecast the dynamics of stocks included in the index basket or belonging to the segment that correlates with the index. Also, the stock index can be used to predict financial crises or sharp surges in the market, which are facilitated by a strong news background, the dynamics of highly volatile stocks, or the creation of a new strong trend in the market.

The history of application of stock indices in the world practice is 129 years old. The first stock index was calculated in 1884 by Charles Dow. He began to calculate the average exchange rate value of 11 companies from the industrial sector, later the number of companies increased to 30, and the industrial index was named the Dow

R. Polyakov (Ed.): EcoSyst 2023, LNNS 1092, pp. 136–144, 2024.
https://doi.org/10.1007/978-3-031-67354-2_16

Jones index. The Dow Jones index is still a benchmark in both foreign and Russian practice. In Russia, stock indices appeared not so long ago, compared to their foreign analogs, they are just beginning their formation. The first Russian index - RTS index was calculated on September 1, 1995 and began to be calculated with 100 points and the calculation began with 25 companies, today it counts 41 companies.

Both domestic and foreign scientists and specialists of various levels were engaged in the study of stock indices. Among foreign scientists we can distinguish Charles Dow, who is the founder of calculations of stock indices, we can also note Henry Varnum Poor, who devoted his life to the study of stock indices and their calculations, in his works reflected the dependence of the stock index on the dynamics of the economy, later founded the companies H.V. and H.W. Poor. Poor. Co., which were engaged in calculating indices and analyzing factors affecting the stock index, as a result of which the companies were transformed into Standard & Poor's with the famous S&P 500 index, which is a reference index in the stock market [1]. Among foreign scientists we can note the works of Y. Fama, who made a significant contribution to the development of stock indices and their dependence on the economy.

Among the domestic scientists who have addressed the issues related to stock indices, we can highlight L.S. Borodavko and V.R. Sorokin, who made a significant contribution to the study of stock indices as indicators of economic movement [2]. We can also note V.I. Lyashenko, who made a great contribution to the consideration of the problems of long-term calculation of the stock index [3]. Rebalancing of the stock index is done once a quarter or once every six months, which, in his opinion, is not enough in the conditions of volatility and high market movement. He puts special emphasis on the years of crisis, saying that in crisis, rebalancing should be done at least once a month because the market movement is very high. But he makes a reservation that the frequency of stock index rebalancing should be correlated with the pace of the crisis.

2 Methodology

The research is organized using traditional methodology. It is about such cognitive tools as analysis, synthesis, induction and deduction, special attention is paid to the causal method, which allows to identify cause-and-effect chains of development of economic events. The practical part of the work, namely the assessment of the degree of sensitivity of stock indices to economic changes, is based on the methodology of econometric analysis, which allowed us to actually confirm the existing regularities of development of different models of the economy.

3 Results

A stock index is an economic indicator that shows the state of the market for securities and other exchange-traded commodities and their trends. It is calculated on the basis of a certain basket of assets (index basket), it can be segmented by industry or by liquidity level (echelons) and so on. The stock index reflects the change in the price of the assets on the basis of which it is calculated and shows their current state. Based on this data, many analysts and investors make a forecast of the stock market movement in one direction or another, so they identify important historical levels, when overcoming which it is possible to talk about the creation of a new strong trend in the stock market. When analyzing the dynamics of the stock index, it is possible to trace how the market reacted to this or that situation. Due to this it will be possible to make forecasts about market reactions in the future [4, 5].

There are a large number of stock indices on the Moscow Exchange. Currently, 29 stock indices are calculated on the Russian stock market. There are only two methods of index calculation used in the Russian Federation: arithmetic mean and weighted arithmetic mean. Of the 29 stock indices, 21 are calculated by weighted arithmetic mean, of which 16 use the price of financial instruments as a weighting indicator, and 5 are weighted by liquidity. The remaining 8 stock indices are calculated by simple arithmetic mean. Calculations of indices by simple geometric mean and weighted geometric mean are not used on the Russian stock market. Economists explain this by the fact that the Russian stock market is just beginning its formation and is very mobile, which is why such calculations are not used, as they will not fully reflect the market dynamics and quickly react to changes or news background.

All stock indices can be divided into 2 large groups: stock indices and bond indices.

In turn, the equity index is divided into main indices, derived stock indices from the Moscow Exchange Index and sectoral indices. All Russian equity indices are calculated as a capitalization-weighted arithmetic average. The construction of equity indices is identical, the index includes highly liquid shares of companies with high capitalization, with the exception of the MosBirch index of medium and small capitalization.

Bond indices are divided into main bond indices and supplemental bond indices. Additional bond indices include bonds that are not included in the main bond indices. The reasons why indices are not included in the main bond indices can be very diverse, for example, they do not correspond to the yield level of the bond, high duration level, etc.

Russian bond indices are calculated by total return of bonds and their net rents. The main indices are calculated in real time and additional indices are calculated once a day.

There are several hundred stock indices in the world stock market. Not all of them are widely popular among investors, but all of them are to a greater or lesser extent indicators of the state of an economic sector or the economy as a whole [6, 7]. There is a relationship between stock indices, they can correlate with each other, thus reacting to certain events in the same way.

Table 1 shows the correlation between the MosBirch index and 26 most popular world stock indices.

The correlation of indices is influenced by many factors. For example, one of the factors of correlation growth is the level of development of trade relations between states. The high correlation between indices should also be considered through the prism of the index basket filling, as this factor has a significant impact. In the Russian indices, the oil and gas industry accounts for the largest share, followed by the financial sector and metallurgy in third place (Fig. 1). All the world indices, having approximately the same index basket filling, will have a high correlation with the MosBirch index.

In addition to the relationship with the world stock indices, the MosBirch index is influenced by external and internal factors. Among the internal factors we can emphasize

Table 1. Closeness of correlation between foreign indices and MosBirch index

Index name	Linear correlation coefficient	Index name	Linear correlation coefficient
Denmark KFX	0.64	Bovespa	0.46
FTSE 100	0.63	AMEX Composite	0.44
AEX	0.62	Nikkei 225	0.44
CAC 40	0.61	IPC	0.43
ISE National 100	0.60	KLSE	0.42
DAX	0.59	BSE 30	0.41
Bel 20	0.58	MerVal	0.40
STI	0.57	NYSE Composite	0.39
HEX	0.57	NASDAQ	0.33
BUX	0.52	S&P500	0.31
SSMI	0.51	DOW JONES	0.31
ASX	0.47	KSE 100	0.30
BIST 20	0.47	FTSE China A50	0.25

Source: compiled by the authors

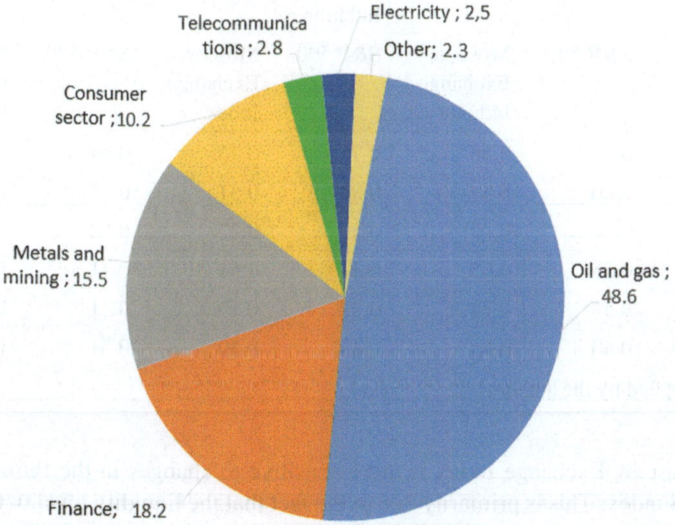

Fig. 1. Structure of Russian main stock indices, %. Source: MosBirch

the refinancing rate, inflation rate and money supply [8]. Among the external factors we can mention the Brent oil rate and the ruble-dollar exchange rate [9].

The study compares the sensitivity to internal and external factors of the MosBirch index and American indices. The choice of American indices was determined by the leading position of the US economy on the world stage.

The assessment of the degree of correlation between Russian and American indices showed the following results. Over the last few years, the degree of correlation has been decreasing due to such reasons as the fall in the beta coefficient of Russian stocks included in the MosBirch index, while the growth of this indicator of U.S. stocks included in the U.S. indices under consideration, the strengthening of trade confrontation between Russia and the U.S., the decreasing dependence of the Russian market on foreign benchmarks.

World stock markets are subject to the influence of internal and external factors. The Russian stock market, as part of the world stock market, is not an exception. The influence of factors can both weaken in certain periods of time and strengthen [10]. The drivers of strengthening or weakening can be various situations in the world stock market, including crises (global, local), the fall of commodity markets, weakening and strengthening of the national currency.

Among the internal factors influencing the Mosbirji index and American indices (the S&P500 index is chosen as a basis for comparison), we can single out the refinancing rate, inflation rate and money supply (Table 2).

Table 2. Influence of internal factors on MosBirch index and S&P 500

Period	refinancing rate		Inflation rate		Money supply	
	S&P 500	Moscow Exchange Index	S&P 500	Moscow Exchange Index	S&P 500	Moscow Exchange Index
2022	0.71	0.86	0.70	0.76	0.64	0.67
2021	0.20	0.51	0.84	0.31	0.57	0.59
2020	0.20	0.49	0.21	0.22	0.62	0.42
2019	0.35	0.65	0.38	0.51	0.41	0.37
2018	0.56	0.74	0.27	0.19	0.31	0.31
over 5 years	0.40	0.65	0.48	0.40	0.51	0.47

Source: compiled by the authors

The Moscow Exchange Index is more sensitive to changes in the refinancing rate than the US index. This is primarily due to the fact that the liquidity level in the Russian market is much lower than in the US market. When the interest rate decreases, there is more available liquidity, which pushes people to invest in riskier assets. It should not be forgotten that interest rates in Russia are much higher than in the US. High interest rates stimulate investments in bank deposits, while low interest rates make this product unattractive for investment.

The sensitivity to changes in inflation rates of the Mosbirji index and the US index is approximately the same. This is due to the high efficiency of the monetary policy of the two countries. A small jump in the sensitivity of the U.S. indices to changes in the rate of inflation occurred due to the acceleration of inflation. In 2022, inflation accelerated to 9.1%, which was a new record for the US.

The Moscow Exchange index has become less sensitive to money supply dynamics from 2020 than the US index. The increase in the sensitivity of US indices from 2020 was due to the "unlimited QE" package offered by Biden. The S&P 500 jumped 71% by 2021. In Russia, however, there is a budget deficit and a decrease in the money supply, so in 2023 the sensitivity of the MosBirch index to the dynamics of the money supply began to decrease.

In addition to internal factors affecting stock indices, there are external factors. These include, for example, the exchange rate of the national currency and the price of oil (Table 3).

Table 3. Influence of external factors on Mosbirji index and S&P 500

Period	Exchange rate		Brent crude oil	
	S&P 500 (USD/EUR)	Moscow Exchange Index (RUB/USD)	S&P 500	Moscow Exchange Index
2022	0.34	0.20	0.34	0.30
2021	0.27	−0.54	0.55	0.24
2020	0.31	−0.59	0.42	0.66
2019	0.30	−0.66	0.24	−0.24
2018	0.32	0.45	0.47	0.80
in 5 years	0.31	−0.23	0.4	0.30

Source: compiled by the authors

The Moscow Exchange Index in 2022 is less sensitive to the dynamics of the national currency than the U.S. index. This is primarily due to the fact that the dollar is the world currency, many transactions are conducted in this currency. The Mosbirzhi index is filled with export-oriented companies, which are more dependent on the dollar exchange rate. But since 2022 the dependence on the dollar has started to decrease and the transition to settlements in the national currency and currencies of friendly countries. American indices are quite sensitive to the dynamics of the USD/EUR currency pair. This is due to the growth of the dollar exchange rate and loss of purchasing power of the dollar in the Eurozone.

The Mosbirji index is also more sensitive to the dynamics of Brent oil than American indices. The influence of oil price on the Russian index is significant, as the oil and gas industry has the largest share in the index. The share of oil and gas companies in the American indices is not so large [11, 12].

The analysis allowed us to state that the Russian index is more sensitive to the influence of external and internal factors. We have to admit that the Russian economy experiences internal and external shocks more often compared to the US economy.

4 Discussion

A stock index performs a number of functions that help analysts and investors assess the picture that exists and is being created in the market.

1. Informational function. A stock index provides an investor or analyst with objective information about the direction of movement of stock exchange quotations, about the presence of a trend in the market, as well as about the speed of development of shares included in the index basket.
2. Forecasting function. Based on historical data of index dynamics, it is possible to make a forecast of future movements, as well as to trace the dynamics and identify turning points of the index, making a conclusion about its reactions to certain situations occurring on the market, in the economy, in the sector, and based on this to build a forecast.
3. Indicative function. The index acts as an indicator of the state of the economy, stock market, separate sector of the economy, separate issuer. That is, the growth of the stock market reflects the growth of the economy, the stock market or a particular sector of the economy.

In its own way, the stock index is a measuring tool. It allows to give an objective assessment of the state and identify the prospects of both the stock market and a particular segment of the economy.

The scope of application of stock indices is very diverse. Foreign experience in the use of stock indices to analyze the stock market and the economy is much greater than in Russia. Over the past 20 years, scientists have conducted many studies related to the nature of stock indices and assessment of their sensitivity to macroeconomic changes. The growth of publications on this topic was observed after the global crisis of 2008, which showed the need to create a system of crisis indicators. Economists divided into two blocks in assessing the role of stock indices.

Some believe that stock indices have one main area of application (the stock market) and a number of secondary ones, while others insist on the multidimensional application of stock indices in assessing certain economic processes.

Proponents of the first point of view believe that the main area of application of stock indices is the securities market [13]. The stock index acts as an indicator when analyzing the securities market conditions, as well as a benchmark when forming an investment portfolio.

Authors adhering to the second point of view argue that the stock index can be used not only in the securities market, but also in the study of any macroeconomic processes. The stock index can be used

- as an indicator reflecting the situation on the stock market and in the economy;
- as a solution to the problem of "statistical incompleteness";
- as an investment vehicle;
- as a basis for forming the structure of the investment portfolio;
- to assess factors affecting changes in the value of securities;
- as an indicator for forecasting the future state of the economy [14, 15].

We are close to the second position. Stock indices have a wide area of application, it is impossible to distinguish their primary and secondary area of application. A stock index is a stock exchange indicator, it is used to assess the stock market conditions, it serves as a basis for creating an investment portfolio, it can also be considered as an independent financial asset. All these spheres of application of the stock index are important, and it would be a mistake to give preference to any one of them.

5 Conclusion

Stock indices are used everywhere, this is due to the fact that they play an important role in the stock market. Indices make it possible to analyze the state of the stock market, to consider historical periods and identify patterns, as well as to identify development trends and make a forecast. The stock index is an indicator of the state of the economy, according to the index you can judge the current state of the country's economy. The stock index also plays an important role for the investor, it helps the investor to assess the state of the investment portfolio.

There may be many reasons for low or high correlation between indices. For example, different methods of calculation, filling of the index basket, foreign policy difficulties between countries, etc.

Since 2020, the MosBirch index has become more sensitive to domestic factors. One of them is the slowdown of inflation in Russia and, consequently, the emergence of available liquidity in the market, with investors investing in riskier assets. The sensitivity of the MosBirch index to changes in the money supply is decreasing, as the Ministry of Finance has forecasted the budget deficit until 2026, so the market has already played back this news and formed important levels. The Russian economy is still dependent on external factors, among which we can mention the price of Brent oil and the exchange rate of the national currency against the US dollar. The commodity export-oriented economic model cannot ignore these exogenous factors.

References

1. Lee, K.: Geopolitical risk and household stock market participation. Financ. Res. Lett. **51**, 103328 (2023). https://doi.org/10.1016/j.frl.2022.103328
2. Borodavko, L.S., Sorokina, V.R.: Impact of economic sanctions on stock indices of different countries. Issues Sustain. Dev. Soc. **9**, 69–79 (2022)
3. Lyashenko, V.I.: Stock Indices of Foreign Markets. Magister, Moscow (2007)
4. Yao, Y., Zhang, Z., Zhao, Y.: Stock index forecasting based on multivariate empirical mode decomposition and temporal convolutional networks. Appl. Soft Comput. **142**, 110356 (2023). https://doi.org/10.1016/j.asoc.2023.110356
5. Wang, Y., Wu, J., Shi, Y.: Stock index prediction using global market indices: a Granger causality-based graph representation learning method. Procedia Comput. Sci. **221**, 797–804 (2023). https://doi.org/10.1016/j.procs.2023.08.053
6. Zhang, X., Li, G., Li, Y., Zou, G., Wu, J.G.: Which is more important in stock market forecasting: attention or sentiment? Int. Rev. Financ. Anal. **89**, 102732 (2023). https://doi.org/10.1016/j.irfa.2023.102732

7. Abuzayed, B., Bouri, E., Al-Fayoumi, N., Jalkh, N.: Systemic risk spillover across global and country stock markets during the COVID-19 pandemic. Econ. Anal. Policy **71**, 180–197 (2021). https://doi.org/10.1016/j.eap.2021.04.010
8. Xu, Y., Wang, J., Chen, Z., Liang, C.: Economic policy uncertainty and stock market returns: new evidence. North Am. J. Econ. Finan. **58**, 101525 (2021). https://doi.org/10.1016/j.najef.2021.101525
9. Alsalman, Z., Herrera, A.M., Rangaraju, S.K.: Oil news shocks and the U.S. stock market. Energy Econ. **126**, 106891 (2023). https://doi.org/10.1016/j.eneco.2023.106891
10. Dong, S., Zheng, J., Jia, H., Zhang, Z.: Impact of capital market internationalization on stock markets: evidence from the inclusion of China A-shares in the MSCI emerging markets index. Res. Int. Bus. Financ. **66**, 101989 (2023). https://doi.org/10.1016/j.ribaf.2023.101989
11. Ahmed, M.Y., Sarkodie, S.A.: Counterfactual shock in energy commodities affects stock market dynamics: evidence from the United States. Resour. Policy **72**, 102083 (2021). https://doi.org/10.1016/j.resourpol.2021.102083
12. Stoupos, N., Kiohos, A.: Energy commodities and advanced stock markets: a post-crisis approach. Resour. Policy **70**, 101887 (2021). https://doi.org/10.1016/j.resourpol.2020.101887
13. Lu, F., Ma, F.: Cross-sectional uncertainty and stock market volatility: new evidence. Financ. Res. Lett. **57**, 104202 (2023). https://doi.org/10.1016/j.frl.2023.104202
14. Shu, M., Song, R., Zhu, W.: The 'COVID' crash of the 2020 U.S. stock market. North Am. J. Econ. Finance **58**, 101497 (2021). https://doi.org/10.1016/j.najef.2021.101497
15. Alamgir, F., Amin, S.B.: The nexus between oil price and stock market: evidence from South Asia. Energy Rep. **7**, 693–703 (2021). https://doi.org/10.1016/j.egyr.2021.01.027

Development of Digital Ecosystems of the Real Estate Market: Russian and Foreign Experience

Galina Gudimenko[1](✉), Viktoria Dorofeeva[1], and Anna Shashkova[2]

[1] Kaliningrad State Technical University, Kaliningrad, Russia
galina.gudimenko@klgtu.ru

[2] Moscow State Institute of International Relations (University) of the Ministry of Foreign Affairs of the Russian Federation, Moscow, Russia

Abstract. The article examines the digital transformation of the real estate markets of Russia and the UAE, related to the introduction of digital technologies and formation of digital ecosystems. The main emphasis is given to the influence of digital economy on the economic growth and development of the real estate market, as well as on the features and driving force of the leading foreign real estate markets. The authors note that over the past decade, the UAE has become a center for the creation of digital products and services and has gained a number of powerful competitive advantages due to the smooth digitalization 2:0. Using the method of comparative analysis, it was shown that Russia has moved to the first level of digitalization, called "patchwork automation". Besides, new trends in the formation of digital ecosystems and their impact on the functioning of the real estate market were analyzed. The study reveals active initiatives related to digitalization, shows the development and intensification of transactions using blockchain, big data, artificial intelligence, virtual and augmented reality and other digital technologies. The main problems and new opportunities of the real estate market presented by digital transformation were described. The authors conclude that it is necessary to create a unified ecosystem of the Russian real estate market and increase the level of its digitalization based on the application of advanced foreign experience.

Keywords: Ecosystem · Digitalization · Digital Technologies · Digital Transformation · Innovative Technologies · Real Estate Market

1 Introduction

Digital technologies are increasingly penetrating into all spheres of our life, creating broad prospects for maximizing the benefits of digital transformation. The evolutionary nature of economic agents interaction, models of their innovative activity and relationships with business environment allow us to determine the digital economy as an ecosystem [1].

The digital ecosystem is able to increase transparency and consistency of data in asset and investment management processes. There is a significant number of business

R. Polyakov (Ed.): EcoSyst 2023, LNNS 1092, pp. 145–153, 2024.
https://doi.org/10.1007/978-3-031-67354-2_17

processes and operations in the real estate market that are subjected to transformation and optimization through the use of digital technologies. It will lead to numerous changes in the behavior of developers, sellers, tenants, agents, real estate owners, i.e., all participants of the real estate market. Innovative technologies in the real estate market can considerably improve the quality of customer communication services and many other indicators of the real estate market development. In other words, the real estate market is becoming more open and transparent.

2 Materials and Methods

The implementation of the research tasks was achieved based on the analysis of statistical data on the dynamics of development and value of real estate in the UAE and Russia in recent years. The methodological potential includes a method of comparative analysis, which allows us to establish the importance of real estate markets for the economies of different countries, taking into account the specific circumstances of their development. When writing the article, depending on the specifics of the tasks being solved, various methods of economic research were used: abstract-logical – in setting the goals and objectives of the study; comparative analysis – in the analysis of ecosystems and technologies of real estate markets; inductive and deductive methods.

3 Results and Discussion

The key driving factors in the formation of digital ecosystems of the real estate market are the following: close cooperation with a buyer, efficient processes and compliance with regulatory requirements, since market participants want unhindered interaction from the moment of opening a real estate object to sales maintenance. Digitalization of the real estate market should be considered as the most important requirement of the market, rather than a tribute to time or a fashion trend. Digitalization acts as an evolutionary process that cannot be resisted.

At one time or another, all participants in the real estate market must go through the process of digitalization and the winners will become those people who will respond faster than others to the request of consumers or will be able to make a transaction in the shortest possible time as transparently as possible. The formation of digital ecosystems can contribute to the intensification of the real estate market digitalization process.

Nowadays, digital technologies are becoming more accessible and gripping for developers, architects, realtors and all other market participants. Ecosystems of the real estate market have been established in many countries and the ecosystem can now be considered as a global trend of its development [2]. In this regard, study of foreign experience in the formation of digital ecosystems in the real estate market seems relevant and timely.

The study of the world experience of the real estate market digitalization demonstrates that one of the most developed countries in terms of the use of innovative digital technologies is the United Arab Emirates. In particular, the ecosystem of the real estate market has been created and is actively functioning in the Emirate of Dubai.

Dubai is comparable to the Moscow real estate market, therefore, it is possible to conduct a comparative analysis in order to borrow best experiences for development of the ecosystem.

Comparative analysis is one of the widely used methods of market research. As a part of comparative analysis, the similarities and differences of the researched subjects or processes are established.

The main requirements for conducting a comparative analysis can be formulated as follows:

1) for comparison there should be chosen objects and processes having something in common, i.e., they must be comparable;
2) the analysis is carried out according to essential criteria.

Compliance with these conditions makes it possible to obtain a real result of comparison while the established differences determine the specifics of the subjects and processes under study. The methodology of comparative analysis is based on the use of analysis, synthesis, analogy, modeling.

The utmost importance of conducting a comparative analysis is to obtain information about compared subjects and processes, general trends of their functioning and development. It will allow critically evaluate the experience of various researched subjects.

The area of Moscow is 2561 sq. km, the area of Dubai city is 2964 sq. km.

12.5 million People live in Moscow, 3.5 million people live in Dubai. At the same time there are approximately one thousand developers and 15 thousand real estate agencies in Moscow, about one thousand developers are operating in Dubai, 13 thousand professional real estate agencies are engaged in buying, selling and renting real estate objects. The price per square meter which in 2022 in both Moscow and Dubai is approximately $2,300 per square meter, as well as the volume of commissioned real estate are comparable.

Despite the pandemic period, the real estate markets in Moscow and Dubai have been relatively stable over the past six years, construction volumes have been increasing and there is no decrease in demand. However, the real estate markets of Moscow and Dubai are strikingly different in terms of digitalization.

In Dubai, very convenient schemes for buying real estate are used. Almost all transactions are made online, the share of such sales is more than 80%, and in the remaining 20% of cases personal contact of parties occur at some stages of the transaction.

The process of purchasing real estate is included in a unified ecosystem of the real estate market. Initially you need to find a suitable object directly through the developer or with the help of a real estate agency certified by the local land Department of Dubai. This scheme established at the legislative level protects a buyer from unscrupulous sellers. If we are talking about the purchase of a commissioned object, the parties draw up two contracts: a preliminary and a Memorandum of Understanding which are an analogue of the Russian contract of sale.

At this stage the buyer makes a deposit, then together with the seller, the developer issues a certificate of no objection, the rights of the new owner are registered in the Dubai Land Department and the entire transaction cost is paid. The buyer receives a certificate of ownership (Title Deed) and becomes the rightful owner of the property.

If an investor purchases an object at the construction stage, the registration procedure is shortened. The parties conclude an initial sale agreement and the buyer can only reissue it to obtain a certificate of ownership when the construction is completed.

All transactions carried out by developers, realtors, developers, banks, legal entities and individuals in the Dubai real estate market are performed with digital technologies in a unified ecosystem. Data on all commissioned and under construction objects are included in a single database of real estate objects, access to which is open to all real estate agencies.

In turn, all real estate agencies have their own customer bases that analyze existing and potential customers in Big Data mode. A person who wants to buy, sell or conduct a lease transaction with a real estate object can contact any agency by their choice. Developers do not work directly with buyers.

The entire set of customer interactions with a real estate agency (the beginning of interaction, ordering, payment, use of real estate objects) is included in the Customer Experience ecosystem. At each stage a client contacts the company through digital channels (sometimes personally) and gains their unique experience.

All market participants use particular devices (computers, adapters, systems, etc.) that are connected to other devices and services via the Internet or another network for data exchange. This process is called the Internet of Things (IoT) [3].

Artificial intelligence (AI) and machine learning (ML) are actively used in the activities of all participants in the Dubai real estate market to simplify the work with big data. This scenario is especially relevant for the promotion of facilities and for the banking sector.

Over the last few years virtual and augmented reality technologies - VR and AR - have been widely introduced in the Dubai real estate market to provide potential buyers with the opportunity to feel like the owner of an object before the purchase, completely immerse themselves in the space of future real estate, experience positive emotions from a virtual inspection of a finished real estate object. The use of VR and AR technologies provides the potential buyer with the effect of full presence with maximum realism. Virtual reality gives a clear understanding of the meter and layout of the object, shows the sufficiency (or insufficiency) of space for a particular buyer, as well as it let see a future view from the windows and the internal infrastructure of the object.

There is a great possibility for any buyer to see the objects in virtual and augmented reality mode both from their smart phone and from the office of the developer or the realtor, using virtual reality glasses for full immersion. Recorded video content can be played on any device – computer, TV, laptop, etc.

Virtual and augmented reality for the sale of real estate in Dubai includes 3D real estate visualizations, animated 3D videos, VR tours; panoramic (360°) photographs, etc.

Almost all developers operating in Dubai with the help of digital technologies provide potential clients with the opportunity to observe the progress of construction in real time. More than 80% of buyers of real estate in Dubai according to the Ax Capital agency [4] noted that the use of virtual and augmented reality technologies helped them make a purchase decision and better understand the layout of the property at the start of construction.

Virtual reality technologies are also used in the training of sales staff and realtors. Dialogue simulations in VR, AR or 2D formats are effectively applied for script development, communication or soft skills development. Besides, VR technologies are implemented in the framework of design tasks, the entire life cycle of real estate for beginning architects, designers and builders.

Real estate agencies widely apply Omni in their work i.e., integrated consumer perception of a product or service through the use of all possible channels that ensures the completeness of the consumer perception process.

Another area of digital ecosystems development in the Dubai real estate market is the maintenance of a register and verification of transactions purity. The use of block chain technologies in the activities of the Land Department has significantly simplified the conduct of real estate transactions in Dubai, increased the speed of registration of real estate rights. The real estate market ecosystem functioning in Dubai is accessible and convenient for all market participants, regardless of their physical location.

The Russian real estate market is at the initial stage of digitalization. The first stage in the formation of digital ecosystems of the Russian real estate market was the electronic registration of real estate objects.

Many of the domestic developers are confident that they have already completed the transition to the "digital side". In fact, more than 70% of the Russian real estate market participants are at the initial level of digitalization. All of them have their own websites, however, more often they are only used for providing the necessary information without using modern digital technologies and mobile applications.

Developers and realtors test various elements of digital technologies and predict the economic effect. At the same time IT companies are ready to cooperate with real estate market participants, considering it as promising.

The Russian real estate market participants have widely mastered basic and relatively simple digital technologies. However, only a few of them are making attempts to reorganize business processes through the introduction of digital technologies. The share of Russian organizations participating in the real estate market that have mastered cloud services is 23%, business process digitalization and management systems (blockchain, Customer Experience, Big Data, IoT, AI) – 12.2%, virtual and augmented reality technologies - 5% [5].

As the world's population is doubled by 2050, the demand for real estate will also double. Citizens of the world will feel more comfortable using technology and their opportunities will expand. Due to interaction with the real estate industry to one degree or another, the lives of many people will be exchanged in the near future.

Currently, the development of digital technologies by participants of the Russian real estate market is conducted in several directions at once [6]. Management companies use the Internet technologies to interact with residents. Individuals are interested in security and video surveillance systems. Developers are implementing individual elements of a "smart home" to increase the attractiveness of their project.

There is no doubt, the driver of digitalization of the real estate market is credit organizations that unlike other market participants use almost all modern digital technologies. Nevertheless, there are practically no digital banks on the Russian market. The reason for this is the specifics of each particular bank, isolation and unwillingness to close interaction of banks within the framework of creating unified ecosystems. IT infrastructures introduced at the initial stages of banks digitalization are characterized by low flexibility and integration difficulty with new digital solutions [7]. The situation has especially worsened with the introduction of sanctions and the need to switch to domestic software.

The main problem of digital systems formation of the Russian real estate market is that no one wants to pay for them: neither buyers, nor developers. Even in the largest segments of the residential market (mass and business class), customers are not willing to pay for additional options. Prices have increased by 15–20% over the past two years and developers are not going to add any new options not to affect the final cost.

As examples of the active use of digital technologies by developers in Russia, we can name the company Aktivo. Due to this company there is an opportunity to invest from 300 thousand rubles in commercial real estate in Russia, "packaged" in a fully digital ZPIF (closed-end mutual investment fund) and the company Investors which sells shares of commercial real estate that brings rental income through blockchain technology. One could also name the company PNK Rental, that invested in industrial real estate but it announced the beginning of the liquidation procedure. The assets that make up the fund will be sold and the received funds will be paid to investors at the end of 2023.

The analysis of active initiatives related to digitalization carried out within the framework of our study indicates the development and intensification of the use of blockchain transactions. This is especially noticeable in the banking sector and in the sector of state registration of the real estate. Rosreestr is one of the first public services to introduce blockchain technologies into their activities where the Moscow Department of Information Technology, the Federal Tax Service, commercial banks, Rostelecom, the Agency for Housing Mortgage Lending, etc. can participate as hubs.

Registration of the real estate using blockchain technologies is convenient for officials, consumers, and developers [8]. Blockchain significantly simplifies and accelerates the introduction and coordination with the Federal Register of information about objects. However, we note that these days all contracts that are registered in digital form are duplicated in paper form.

The practice of forming digital ecosystems in the Russian market of immobility differs from similar practices in other economically developed countries. Active digitalization of processes can be observed only in the largest realtors and in large banks in Russia. The rest of the Russian market participants are not in a hurry to join digital technologies yet.

A study of the websites of the largest Moscow developers allowed us to conclude that they are all built on a similar principle: on the main page, the real estate objects are differentiated into groups: new buildings, business class, mortgage, market (related products and services), business premises, etc. When selecting the section of interest, a set of objects offers opens, i.e., delivery date, location on the map and the plan of the object. Besides, for individual objects, the possibility of conducting a transaction remotely (electronically) is additionally indicated.

If we analyze the websites of the largest Moscow real estate agencies, a similar picture is observed: information about objects, location on the map, infrastructure, the possibility and conditions of a mortgage and several photos are presented. Unfortunately, it is not yet possible to talk about the use of a full range of modern digital technologies by Moscow developers, realtors and other participants in the real estate market, as well as about the availability of digital ecosystems.

In Moscow about 150 thousand agencies, various organizations and real estate specialists provide real estate services such as individual entrepreneurs, large and small organizations, individuals who do not have the status of an entrepreneur but they consider themselves as specialists. According to this amount, about 15 thousand are represented by real estate agencies. Due to this differentiation of the Moscow real estate community, it is possible to talk about the digitalization of realtors' activities with a high degree of skepticism because digital technologies are introduced only by large progressive agencies.

The problem of using the possibilities of digital technologies is especially acute in the secondary real estate market. For instance, when selling a real estate object, it is not enough to place a description of the object, photo and video materials on a digital board or platform. It is also necessary to get a lot of certificates and documents from the management company (when selling a residential property), from organizations that supply resources: light, heat, etc., and it is impossible to get the necessary certificates remotely.

Participants of the domestic real estate market are committed to the old traditions that is making real estate transactions, you should hold several meetings, evaluate the object during a personal inspection, manually collect information to check the counterparty, analyze the information available in open sources, personally check all documents certifying ownership or the right to make a transaction. At the same time completely different business scenarios are being used in other countries, new information technologies are being introduced.

It is supposed the most acute problem of ecosystem formation of the Russian real estate market to be disunity of market entities. Each relatively large developer, bank or real estate agency creates their own ecosystem and conducts operations there, for instance, real estate management, design, search and supply of construction equipment and materials, etc.

There are currently more than fifty digital banking ecosystems and more than thirty corporate ecosystems of real estate developers operating in the Russian real estate market. However, these ecosystems and platforms do not allow integrating different databases with each other. A lot of available services and digital platforms are used either in individual real estate transactions or accompany transactions from start to finish but in limited market segments. To solve the problems of the Russian real estate market it is necessary:

- to start work on creating a unified digital platform for the real estate market, VTB digital services can be taken as a basis, since nowadays it is the widest platform with a large set of functions. The development and control of the functioning of the platform can be entrusted to the newly created Industrial Competence Center under the Ministry of Construction of the Russian Federation;

- create a website constructer from the available services in the shortest possible time, following the example of a website constructer for online stores which will make it easier for small companies to create websites with modern services. As a basis you can take the

Tilda website constructer, the Profitbase catalog, personal account, document recognition service, mortgage service, individual products to accompany a digital transaction;

- to study and use foreign experience in the formation of digital ecosystems. We believe that the Russian real estate market does not need to "reinvent the wheel", it is necessary to take from foreign experience all the best that has already been tested, implemented and effectively operates in the real estate market.

It is possible to determine the level of digital ecosystems development of the real estate market on the basis of the digitalization criteria presented in Table 1.

Table 1. The level of digitalization and their characteristics

The level of digitalization	Features of digitalization level
Zero level – analog approach	Numerous visits to the MFC, availability of paper documents, search and visit of real estate objects, difficult identification of developers (sellers), filling out bank questionnaires with a pen, opening accounts during personal visit of the bank
The first level is patchwork automation	Availability of various services for digital mortgages and electronic registration, opening accounts during personal visit of the bank, mismatch of accesses, databases, cabinets
The second level is seamless digitalization	A unified service for all stages of buying and selling real estate, a minimum of paper situations, the advantage of the seller over the buyer in transactions
The third level is seamless digitalization 2:0	The advantage of the buyer over sellers in transactions, entry into the cabinet and control of transactions from any device, legislative support for the digitalization of the real estate market, state support for digitalization and digital ecosystems

Made on the basis of [9].

Taking into account the characteristic features of digitalization levels, it can be definitely stated that Russia having passed the zero level, has moved to the first level of digitalization - patchwork digitalization. There are diverse services for digital mortgages and electronic registration in the real estate market, individual operations are carried out remotely while there is a mismatch of accesses, databases, offices for various market participants. Ecosystems in the Russian real estate market are not formed. Nevertheless, the level of digitalization of the UAE is rated by us as the highest, the third level, or "seamless digitalization 2:0". The characteristics of this level are the advantage of the buyer over sellers in transactions, the access to the cabinet and control of transactions from any device, etc. The ecosystem of the UAE real estate market is convenient for use by all market participants.

Increasing digitalization level of the Russian real estate market based on the use of advanced foreign experience will contribute to the development of the market, as well as solving the tasks of the state in providing citizens with affordable and comfortable housing and ultimately improving the quality of life of the population.

Advantages and disadvantages of digital technologies in real estate cannot be correlated. However, it is obvious that there are much more advantages. Digital technologies are simple, affordable and convenient. In the future functional opportunities that provide digital technologies will only expand. For the development of the Russian real estate market, it is extremely useful to study advanced foreign experience in the formation of digital ecosystems. We believe that the Russian real estate market does not need to "reinvent the wheel", it is only necessary to adopt from foreign experience all the best that has already been tested, implemented and effectively used in the real estate market.

References

1. Dorofeeva, V.V.: Development of the digital economy in terms of new global challenges. In: Polyakov, R. (eds.) Ecosystems Without Borders. Lecture Notes in Networks and Systems, vol. 474. Springer, Cham (2021). https://doi.org/10.1007/978-3-031-05778-6_16
2. Cavallo, A., Ghezzi, A., Balocco, R.: Entrepreneurial ecosystem research: present debates and future directions. Int. Entrepreneurship Manage. J. **15**(3), 1–24 (2018). https://doi.org/10.1007/s11365-018-0526-3
3. Finch, A.: Techwatch: impact of artificial intelligence on real estate. https://www.rodschinson.com/en/insights/techreal-estate/238-impact-of-artificial-intelligence-onreal-estate. Accessed 20 August 2023
4. Axcapital Homepage. https://www.axcapital.ae/ru/buy/dubai/properties-for-sale?utm_source=yandex&utm_medium=cpc&utm_campaign. Accessed 02 September 2023
5. Abdrakhmanova, G., Vishnevsky, K., Gokhberg, L. What is the digital economy? Trends, competencies, measurement H-80: dokl. to the XXII Apr. In: International Scientific Conference on Problems of Economic and Social Development, Moscow, 9–12 Apr. 2022 / National research. Higher School of Economics Univ., Moscow: Publishing House of the Higher School of Economics (2022)
6. Ivanov, M.V., Rumyantseva, S.: The new ecosystem of digital real estate: stages of development, technologies and prospects News of universities. Investment. Construction. Realty. **10**(4), 524–533 (2020)
7. Golaydo, I., Parshutina, I., Cudimenko, G., Lazarenko, A., Shelepina, N. Evaluation, forecasting and management of the investment potential of the territory. J. Appl. Econ. Sci. **8**(2, 48), 8–25 (2017)
8. Grant, P., Putzier, K. Commercial property joins tech revolution as spending soars. Wall Street J. https://www.wsj.com/articles/commercialproperty-joins-tech-revolution-as-spending-soars11562068801. Accessed 12 September 2023
9. Lepinskikh, B.: From the "Stone Age" to a seamless process: how digitalization is changing the real estate market. 2020. https://www.cossa.ru/special/realty/264325. Accessed 04 September 2023

Communication Ecosystem Construction: Comparative Analysis of Communication Strategies of Municipal Securities Issuers

Andrey Koshkin(✉), Matiar Rakhman Khashimi, Maksim Shcheglov, Alina Koxegenova, and Aleksandr Bukhantsov

St. Petersburg State University, St. Petersburg, Russian Federation
a.koshkin@spbu.ru

Abstract. This article tries to attempt a comparative analysis of the communication strategies of municipal issuers in Russia, the USA, Great Britain and Germany. Primary attention is paid to the issue of developing recommendations for improving the communication ecosystem of the process of issuing municipal securities in Russia. The attention of the authors is focused on the interpretation of the concept of "municipal securities" and the features of their issue in the mentioned countries. At the macro-theoretical level, the article is based on the theory of communicative action by J. Habermas and the concept of problem management by K. Tucker and G. Broom. The article also provides an overview of Russian and foreign researches on the issues of the study. The most promising methods for improving the communication ecosystem of the process of issuing municipal securities in Russia were identified: explanatory work with the leadership of municipalities, the creation of the Agency for Municipal Finance of Local Authorities following the example of Great Britain or Sweden, and the creation of a new type of the Individual Investment Account (along with the already existing), aimed at stimulating the acquisition of municipal securities.

Keywords: Municipal Securities · Communication Strategies · Securities Issue · Communication Ecosystems

1 Introduction

One of the most discussed problems in the search for urban development tools is the insufficient provision of the municipal budget. We exist in a world where resources are always in short supply, and that is why they have economic value. In this context, the topic of municipal securities, as a probable source of resources, seems very relevant.

Municipal securities (MS) make it possible to finance the municipal budget deficit, attract external investments, and also mobilize intra-city financial resources of individuals and organizations that are not banks. However, it must be recognized that along with the listed advantages of this tool, it has very serious disadvantages. Perhaps the main thing is the fact that the funds raised in this way will have to be returned, and with interest. The context of the relatively recent wave of municipal defaults in the United

R. Polyakov (Ed.): EcoSyst 2023, LNNS 1092, pp. 154–161, 2024.
https://doi.org/10.1007/978-3-031-67354-2_18

States (and by September 2022, the number of first defaults under the MS was $984.2 million [1]) makes the management problem of this instrument quite clear. When and under what conditions is the use of the issue of municipal securities a justified step? What problems does such a tool help solve, and what problems can it only aggravate?

In addition, it seems reasonable to focus on communication strategies when issuing interbank securities. Therefore, in the legal field of Russia, the USA, Germany, France, etc. interpretations of the very concept of "municipal securities" are very different from each other. The key thing, which will be discussed below, is that the concept of "municipal securities" may also include securities issued by regional governments and legislatures (example: federal subjects in the Russian Federation). The level of attention paid to them by private investors, bank capital and other financial institutions may depend on how issuers position and "communicate" their securities. Also, the communication ecosystem around the issuance process makes it clear the purpose of raising funds. The money received during the issue of interbank securities can be used in a targeted manner to improve infrastructure, subsidize the activities of a municipal enterprise, or simply finance the budget deficit.

Thus, the purpose and objectives of the study can be identified. The purpose of this research is to formulate recommendations for the use of municipal securities emissions as a tool for urban development. The main objectives include:

1) Analysis of different approaches to the interpretation of the concept of "municipal securities" in different countries;
2) Description of the practice of issuing municipal securities;
3) Generalization of the studied practices and formulation of the main conclusions in the context of the application of the studied tools in Russia.

2 Materials and Methods

The methodological basis of the study was the work of S. Glazkov on the analysis of the securities market in Russia (and, especially, the issues of "municipal bonds" conceptualization) [2], as well as the article by E.M. Namsaraeva, where strong arguments are given in favor of the distinction between subfederal and municipal securities [3], their importance will be discussed in the main part of the study. From a macro-theoretical point of view, this work is based on the concepts of problem management and analyzes the research problem posed by working with cases of application of the selected management decision (namely the issue of interbank securities) by municipalities in different countries. The applicability and high effectiveness of problem management in the context of minimizing risk and increasing favorable policy opportunities of government agencies was substantiated by K. Tucker and G. Broome [4]. Analysis of foreign and domestic practices will allow for a more substantive discussion about the problems and prospects of the management decision being studied. During the prosses of using securities analysis, our attention was paid to three parameters: the similarity of interpretations of the concept of securities in regulations, the volume of turnover of the securities market, as well as the uniqueness of individual practices of issuing securities.

With regard to interpretations of communication and the communication ecosystem, the authors rely on the theory of communicative action by Jürgen Habermas [5, 6] and its modern interpretations [7–9].

3 Results

First of all, we have to admit that the research of the identified problem in the domestic literature is not at a high enough level. It is presented rather in the form of separate publications. Such publications, unfortunately, are often dated back to the second half of the 2000s or the first half of the 1900s (although individual authors continue to stand out today), or are not always executed with high quality. Thus, a whole line of publications analyzing the state of the securities market in Russia interprets municipal securities as "debt securities issued by city, regional, and republican entities" [10], which directly contradicts the Budget Code of the Russian Federation, where a clear definition is given: "Municipal securities are securities issued on behalf of a municipal entity" [11]. In other words, in Russia, securities of the constituent entities of the federation are not recognized as municipal securities, but are considered government securities along with government securities of the Russian Federation.

This brings us to the first task of the study - the analysis of different approaches to the interpretation of the concept of interbank market in different countries. There is no unanimity among countries in this regard. The countries that are closest to Russia in terms of the interpretation of the MS are the UK, France, Australia, Sweden and Finland, where they even have their own Agencies for municipal financing of local authorities. In Great Britain this body is relatively new. The process of its formation lasted from 2014 to 2019.

An intermediate interpretation of the MS is in Germany. It views MS as fixed income securities issued to the market by mortgage bankers or public sector lending institutions on behalf of and in agreement with a particular municipality. In this case, the placement itself must be agreed upon with the president of the federal state where the municipality is located.

A diametrically opposite interpretation of the MS contains the US legislation. They are interpreted as "debt securities issued by states, cities, counties, and other government entities to finance day-to-day obligations and finance capital projects such as the construction of schools, highways, or sewer systems." In this case, it is precisely correct to talk about interbank securities as securities issued by either a municipality or a region (state).

These features are important to take into account during a comparative assessment of the securities markets' volumes in the mentioned countries. Moreover, it can be stated that the approach to the interpretation of the MS chosen in the USA is completely inapplicable to unitary countries due to the often-missing intermediate level of management between municipal and national authorities. On this basis, it seems more reasonable to adhere to the interpretation approach of the MS that is set out in the Budget Code of the Russian Federation.

We focus on the analysis of the direct practice of issuing interbank securities. It seems reasonable to start with Russia.

Unfortunately, searching for information about the MS in Russia is difficult. Various specialized sites duplicate errors from some domestic studies and sites for monitoring prices and profitability of securities and reflect their government securities of the constituent entities of the federation [12]. There is not a single MS in the list presented there.

Nevertheless, decisions on the issue of municipal securities are posted on the websites of city administrations, which allows us to draw conclusions about the current state of the municipal securities market in Russia. Thus, the modern locomotives for issuing municipal securities are Krasnodar [13] and Novosibirsk [14], which were able to attract 3 and 5 billion rubles, respectively.

Some data up to 2015, allow us to analyze the state of the securities market during this period [15]. The issues were constructed according to one type, the bonds were documentary, bearer, with a constant coupon income, debt amortization and mandatory centralized storage of the global certificate of issue. Also, over four years (from 2011 to 2014), the Ministry of Finance of the Russian Federation registered the terms of issue of 20 issues of 8 cities: Krasnoyarsk (two issues), Novosibirsk (eight issues), Volgograd (three issues), Krasnodar, Kazan, Tomsk (three issues), Omsk, Volzhsky, Volgograd region. The total amount of securities issued during this period amounted to 16.1 billion rubles.

Unfortunately, based on the analysis of modern practices, the conclusion from the 2015 studies can be confirmed. The MS is still practically not used for the socio-economic development of cities, but serves only as means of covering the deficit of city budgets. Also, the main holders are various organizations, and not individuals living in the cities. We can also note a decrease in the activity of Russian municipalities in issuing their own securities.

An analysis of the international practices of issuing interbank securities in the United States seems important, since intermediary securities are the most developed in this country. It is necessary to make a reservation that in the US both municipal securities and subfederal securities (states) are recorded as securities. However, one way or another, according to the Federal Reserve, up to 73% of municipal bonds are in the hands of individual private investors - directly or through investment funds. About 12% are assets in the portfolios of insurance companies, about 11% are in the portfolios of commercial banks, savings banks and investment banks.

It should also be noted the wide variety types of long-term obligations of interbank securities in the USA [16]. The following types can be distinguished:

- *Full faith and credit pledge* are obligations whose repayment is secured by all the income of municipalities;
- *Limited guarantee bonds* are obligations whose repayment is secured only by certain items of tax revenue or fees;
- *Moral guarantees* are a series of bonds, payments for which are secured by the "moral obligations" of municipalities. Despite the absence of legal obligations, it is expected that in the event of non-payment, the authorities will seek budgetary funds to repay such securities;
- *Revenue bonds* are obligations that are repaid from the income of a certain enterprise, facility or project and do not have a guarantee from the municipal authorities. Funds received as a result of such bonds placement are spent on projects related to certain infrastructure facilities - bridges, roads, water treatment systems, and other objects that require significant capital investments.

The next country that seems reasonable to pay attention to is the UK with its practice of the Municipal Finance Agency for the Local Governments [17]. It monitors the unreasonable use of securities, provides methodological and organizational support for the placement of securities on the market. The Agency also promotes the dissemination of practices for the reasonable use of securities. According to the studies from the University of Sheffield, this significantly reduces capital costs for long-term financing of infrastructure projects in the UK [18]. It seems that this management decision fully corresponds to the problems of the necessary institutional framework for external control over the issue of securities, posed in the domestic scientific literature [19].

Another country whose experience seems fundamentally important for the analysis is Germany. German states are important players in the European capital market. Thus, in December 2016, the number of outstanding securities (together with subfederal securities) amounted to 364 billion euros. This is more than seven times higher than in Spain (52 billion euros). Italy and France were able to attract 22 and 18 billion with this tool, respectively [20].

Researchers highlight several aspects of this success:

- Germany maintains its AAA credit rating. This is the only euro area state that maintained its highest performance during the sovereign debt crisis and up to now;
- Existing mechanisms of mutual support and financial equalization of federal lands and municipalities in combination with the principle of federal solidarity;
- Favorable regulatory environment. Issued securities have a risk weight of approximately zero and a high credit rating;
- A significant share of involved insurance companies, funds, banks and even central banks of other countries. This is a result of the need to meet strict liquidity criteria in investments (including LCR regulation).

Of particular interest are the MS that are issued jointly by cities. Thus, in 2013, the cities of Nuremberg and Würzburg issued joint municipal bonds for the first time. This model was quickly recognized as successful, and this practice spread throughout the municipalities of North Rhine-Westphalia. In February 2014, two more groups of similar issuers were registered. Mainz and Ludwigshafen also showed particular activity in this direction.

4 Discussions

At the moment of preparing this article, the Russian securities market is experiencing a decline. The long-term practice of using MS as a tool to cover the city budget deficit in the context of an increasingly centralizing model of fiscal federalism naturally leads to the replacement of MS emissions with interbudgetary transfers that are more convenient in this regard. Particularly noteworthy is the significant expansion of the "Other" line in interbudgetary transfers during the COVID-19 pandemic [21].

The US system of targeted securities seems to be extremely developed: from financing multifamily housing developments to reducing the potential environmental impacts of commercial activities. Coupled with serious tax benefits that are provided for investing in securities, this makes it possible to attract individual private investors to the securities market. Such tools allow us to talk about the genuine use of MS emissions as a tool for managing urban development.

However, we cannot ignore the experience of Germany, where we highlight the special role of horizontal inter-budgetary connections, which is reflected in the risk weight coefficient and credit ratings of the MS. This practice encourages more conservative financial institutions such as banks to invest in urban development. Their interest lies in meeting strict liquidity criteria in investments.

These results and recommendations are fully consistent with the research of Professors V. Pashkus [22–24], M. Livingston [25], D. Cestau, B. Hollifield, Dan Li and N. Schürhoff [26]. In these studies, the authors note the reality and systemic nature of the shortage of funds from municipal budgets in the Russian Federation with the growth of mandatory spending provided for by the institutionalized division of powers to ensure government functions in the regions of the Russian Federation. In addition, foreign researchers see the particular effectiveness of tax incentives for the development of the domestic market for municipal securities [27].

We also would like to draw attention to the whole direction of publications analyzing the state of the interbank securities market in Russia, which interprets municipal securities as "debt securities issued by city regional, regional, republican entities" [5], which is considered as fundamentally incorrect in analysis of the current communication ecosystem for issuing municipal securities in Russia. The reason for such disagreement is the direct contradiction of this interpretation to the Budget Code of the Russian Federation, which creates significant distortions in the analysis of empirical material.

5 Conclusion

Practical recommendations for improving the tools of communication strategies for issuing securities in Russia can serve as conclusions for this article.

Firstly, *explanatory work with the leadership of municipalities.* City managers need to be introduced to the practices of the municipal securities and successful cases of applying this tool. The main goal of this paragraph is to overcome the perception of the interbank market, which has developed in the Russian municipal management tradition, solely as a way to cover the city budget deficit.

Secondly, *the creation of an Agency for Municipal Financing of Local Governments*, following the example of Great Britain or Sweden. Such an agency would promote the wise use of municipal securities and monitor their application. They can also provide methodological assistance in placing securities on the financial market. Another function of this agency can include educational activities in relation to the general public about the differences between municipal securities and subfederal (regional) securities.

In addition, this agency can promote the development of horizontal budgetary connections of municipalities following the example of Germany. It is not necessary to immediately implement a joint issue of interbank bonds by several municipalities, but such a practice of more responsible budget equalization can compensate for the negative effects of Russia's over-centralized model of fiscal federalism.

Finally, adopting the US experience, we can consider the possibility of *creating a new type of Individual Investment Account* (along with the existing one). This type could focus specifically on providing tax incentives for citizens who decide to invest in municipal securities. Additionally, it makes sense to explore the possibility of targeted issuance of interbank securities for specific development tasks of the municipality. All this will help attract citizens' funds to the securities market. It is worth noting the willingness of Russian citizens to invest in the financial market. The number of registered individual private investors in Russia is growing at a significant pace and the emergence of a virtually new investment instrument looks quite natural.

References

1. Miller, J.V.: Municipal bonds: hints of recovery emerge. Official website of the CFA Institute. Nuveen (Electronic resource). https://clck.ru/33ZNHG. Accessed 16 January 2024
2. Glazkov, S.: Municipal bonds in the Russian Federation. Stocks Bods Market **6**, 67–70 (2011)
3. Namsaraeva, E.M.: Subfederal and municipal securities. Financ. Credit **16**(88), 23–35 (2001)
4. Tucker, K., Broom, G.: Managing issues acts as bridge to strategic planning. Public Relat. J. **11**(49), 38–40 (1993)
5. Habermas, J.: Theorie des kommunikativen Handelns. B.I. Suhrkamp, Frankfurt am Main (1987)
6. Habermas, J.: Moral Consciousness and Communicative Action. MIT Press, Cambridge (1990)
7. Patberg, M.: Supranational constitutional politics and the method of rational reconstruction. Philos. Soc. Criticism **40**(6), 501–521 (2014)
8. Voronin, V.A.: Public institutions and economic development: a study of the relationship on the example of Latin America, Africa and Asia. Bulletin of the SNO of St. Petersburg University, Issue **1** (2024)
9. Pedersen, J.: Habermas' method: rational reconstruction. Philos. Soc. Sci. **38**(4), 457–485 (2008)
10. Simonov, E.S., Baranova E.S.: Municipal securities as a special type of securities. In Gonin, V. (eds.) All-Russian Scientific and Practical Conference 2019 April 29, pp. 102-108. Zabaykal'skiy gosudarstvennyy universitet, Chita (2019)
11. The government of the Russian Federation web-site. http://government.ru/docs/all/96557/. Accessed 21 January 2024
12. Website of the information agency "Banki.ru". https://www.banki.ru/investment/search/sector-mun/. Accessed 16 January 2024
13. Website of the Krasnoyarsk city administration. http://www.admkrsk.ru/citytoday/finance/Documents/Mun_obligatsii/Постановление%20№%20817%20от%2013.10.2020.pdf. Accessed 16 January 2024
14. Website of the Novosibirsk city administration. https://minfin.gov.ru/common/upload/library/2020/03/main/Otchet_ob_itogakh_emissii_gosudarstvennykh_tsennykh_bumag_goroda_Novosibirska_za_2019_god.pdf. Accessed 16 January 2024
15. Edronova, V.N.: Experience and features of bond loans of Russian cities at the end of the 20th - beginning of the 21st centuries. Financ. Credit **31**(655), 23–36 (2015)
16. Official website of the US government. Section: "Securities Commission". Subsection "Municipal securities". https://www.investor.gov/introduction-investing/investing-basics/investment-products/bonds-or-fixed-income-products-0. Accessed 16 January 2024
17. Official website of the UK Council Finance Agency. https://ukmba.org/. Accessed 16 January 2024
18. Shamseen, R.: The Application of Municipal Bonds to fund Local authorities projects in England; Barriers and Solutions. Diss. Universityof Sheffield (2022)
19. Fomin, M.V., Lakhno, Y., Pyshnograi, A.P.: Munitsipalnye ob-ligatsii kak instryment razvitiya infrastruktury [municipal bonds as a tool for development of infrastructure]. Public Admin. Issues **2**, 185–210 (2019)

20. Handbook of German States "Key players in European capital markets". https://www.research.unicredit.eu/DocsKey/credit_docs_2016_157740.ashx?M=D&R=41841879. Accessed 16 January 2024
21. Koshkin, A.V.: The economic aspect of Russian federalism: interbudgetary transfers during the onset of the COVID-19 pandemic Marketing MBA. Mark. Enterp. Manage. **1**(12), 69–81 (2021)
22. Pashkus, V.Y., et al.: Realization of breakthrough positioning strategy for agritourist objects. J. Environ. Manage. Tourism **3**(15), 439–448 (2016)
23. Pashkus, V., Pashkus, N.A., Krasnikova, T.S., Asadulaev, A.B.: Estimating a university brand in the new economic conditions: concept, principles, technique. Int. J. Econ. Financ. Issues **3S**(5), 100–104 (2015)
24. Pashkus, V., Pashkus, N.A., Shvetz, P.V.: Competitiveness of regional healthcare institutions in the era of COVID-19. Problemy sovremennoy ekonomiki **1**(81), 140–144 (2022)
25. Livingston, M.: The pricing of municipal bonds. J. Financ. Quant. Anal. **2**(17), 179–193 (1982)
26. Cestau, D., Hollifield, B., Li, D., Schürhoff, N.: Municipal bond markets. Annu. Rev. Financ. Econ. **11**, 65–84 (2019)
27. Drukker, A.J., Gayer, T., Gold, A.K.: Tax-exempt municipal bonds and the financing of professional sports stadiums. Natl. Tax J. **73**(1), 157–196 (2020)

Innovation Ecosystems

Ensuring Sustainable Development of the Food Packaging Industry

Varvara Altunina[1], Valery Voytyuk[2], and Nadezhda Li[1](✉)

[1] Immanuel Kant Baltic Federal University, Kaliningrad, Russia
nli124127@gmail.com
[2] Kaliningrad branch of the Russian University of Cooperation, Kaliningrad, Russia

Abstract. The paper addresses the relevance of sustainable development of socio-economic systems. The main stakeholders and their roles, methods and tools to ensure meso and micro levels of sustainable development are studied. Micro-decomposition of sustainable development goals makes rational implementation of separate development projects in individual industries and enterprises. This approach not only ensures consistent project implementation, but also enables identification, analysis, assessment and mitigation of risks generated by the projects. Particular focus is placed on feasibility of the projects oriented towards one of the goals of sustainable development – sustainable consumption and production in the food packaging industry. The global food packaging market demand is showing a rapid growth trend. The sustainable consumption and production network include governments, civil society, the private sector, and academic institution and international organizations. The interaction of key stakeholders to ensure sustainable development in the eco-friendly (sustainable) food packaging industry is analyzed. Sustainable packaging involves materials with the minimum environmental impact throughout its life cycle. The production of such packaging is aimed at reducing waste, saving resources and promoting renewable materials.

Keywords: Sustainable Development · ESG Principles · Environmental Impact · Food Packaging · Risk Analysis · Development Projects

1 Introduction

The issues of the harmonious relationship between nature and humans are of current relevance. Crisis events of recent years have catalyzed research into the quality of modern economic growth and enhanced the search for innovative development models to ensure sustainable development of ecosystems.

According to the UN Development Goals Report (Special Edition), the COVID-19 pandemic and a triple crisis of climate change, biodiversity loss and pollution give rise to long-term devastating effects [1].

The global situation is getting worse year on year due to growing production and increasing material consumption. Increased material extraction reduced global circularity: from 9.2% in 2018 to 8.6% in 2020 and to 7.2% in 2023. This means that materials

R. Polyakov (Ed.): EcoSyst 2023, LNNS 1092, pp. 165–174, 2024.
https://doi.org/10.1007/978-3-031-67354-2_19

that return to the global economy after their useful life, also known as recycled materials, account for 7.2% of all material inputs into the economy [2].

Sustainable development is crucial for all countries, as confirmed by the studies as well as global actions taken and measures included in the national policies of the countries.

In 2015, the UN summit set 17 sustainable development goals enshrined in the UN declaration transforming Our World: the 2030 Agenda for Sustainable Development. In total, 193 countries have committed themselves to achieving sustainable development goals.

The relevance of sustainable development is recognized not only at the international and government levels, but also by business, end consumers, and educational and research institutions. Thus, it is obvious that sustainable development can be achieved through coordinated actions of all stakeholders aimed at a common goal.

2 Materials and Methods

The issues of on the need to implement projects aimed at sustainable development goals bring us to Edwards Deming who states the true goals of the company: "However, the true goal of our organization should be its continuous improvement, which will include all aspects of its life (that is, improvement of our products and services, improvement of our relationships with customers and suppliers, improvement of our relationships with society, improvement of the lives of our employees, etc.)" [3].

Micro-decomposition of sustainable development goals makes rational implementation of separate development projects in individual industries and enterprises.

This approach ensures consistent implementation of these projects, and enables identification, analysis, assessment and mitigation of the risks generated by these projects.

The environmental component of these projects is of particular relevance due to an objective need to reduce the environmental load caused by human impact. Economic activities related to generation of various types of energy significantly enhance this impact. Energy transition is one of the most crucial tools to attain the environmental goals of sustainable development. The feasibility of this transition is confirmed by the study conducted at the Fraunhofer Institute for Solar Energy Systems ISE, which reported that renewable energy sources generated about 60% of the total German electricity in 2023 [4].

Today, more and more companies worldwide are underway to integrate ESG risks into their activities, initiatives or strategies. Karen Baum, the sustainability and ESG advisory services leader at BDO USA, PC, notes in her paper: "Despite prevailing attitudes, every organization has some degree of ESG risk exposure." [5].

Much work is being done to reduce waste generation in EU countries. The study results reported by the European Environment Agency [6] "Tracking progress in waste prevention – Descriptive monitoring of waste prevention framework at EU level" (2023) revealed the main trend of annual waste growth of 7.2% in the EU territory. In 2020, joint efforts of 27 countries attained significant results and declined this indicator to the level of 2016 (Fig. 1).

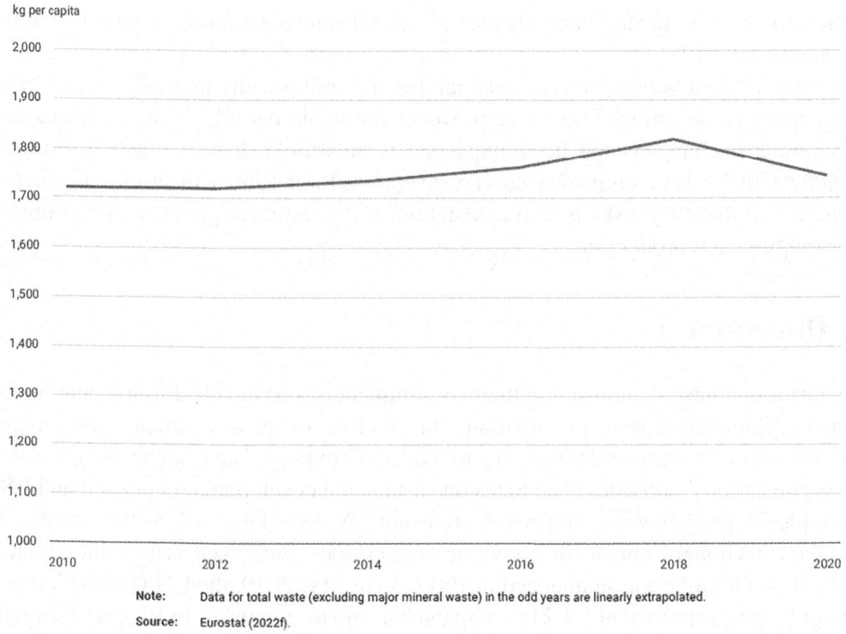

Fig. 1. Total amount of waste, excluding basic mineral waste generated per capita in the EU-27, 2010–2020.

Development projects aimed at sustainable consumption and production in the food packaging industry are proposed to be implemented based on the method of a standard process model of project management [7], which includes five groups of processes:

- project initiation processes;
- planning processes;
- execution processes;
- monitoring and management processes;
- project closure processes, including a comprehensive analysis of the attained goals.

3 Results

We suggest that the key driver of sustainable development projects aimed at sustainable consumption and production in the food packaging industry is adoption of advanced innovative technologies for its production. The country, which has committed itself to principles of sustainable development, must give extensive support to companies that develop new environmental products and facilitate the transfer to relevant innovative technologies. This assertion is supported by the theses from the report by the UN Secretary-General at the 78th Session of the UN General Assembly on the adoption of technological advances and innovation for sustainable development [8].

China is a case in point of the policy to stimulate innovative development processes. The country's leadership has enshrined the commitment of companies to increase the share of expenses on R&D and allocate 5% of sales revenue for these purposes at the legislative level [9].

From a business perspective, sustainability has traditionally implied cost reduction and capital preservation. This strategy works for stable periods. However, instability and long-term change is now becoming a new normal for business. Under conditions of prolonged instability, companies must create options for risk diversification. An effective approach to diversify risks is implementation of development projects on sustainable consumption and production.

4 Discussion

Countries that have demonstrated their commitment to sustainable development implement development projects into their internal strategic and program documents. Organizations follow ESG principles that affect social, environmental and managerial aspects of the organizations' activities. The focus on sustainable development in general and ESG principles in particular is confirmed by relevant UN decisions, such as the initiative to join the UN Global Compact on corporate social responsibility and sustainable development. This initiative was announced in 2000. As of May 2019, the UN Global Compact brought together more than 21,218 companies from 162 countries. In Russia, 22 organizations joined the initiative over the period of just from 2019 (57 units) to 2022 (79 units) [10]. When choosing goods and services, the end consumer evaluates their environmental friendliness. Scientific research conducted at universities is aimed at technologies developed to meet the current needs of business and society as a whole. End consumers opt for eco-friendly products.

However, no scheme has yet been elaborated for the interaction of key stakeholders to achieve sustainable development of the systems.

Consider the interaction of key stakeholders to ensure sustainable development in the eco-friendly (sustainable) food packaging industry.

Sustainable packaging involves materials with the minimum environmental impact throughout its life cycle. The production of such packaging is aimed at reducing waste, saving resources and promoting renewable materials.

The study conducted by the analytical agency Market.US [11] showed a growing trend in the volume of the global food packaging market. By 2032, it will be about US$592.8 billion, up US$358.30 billion from 2022, grown on average by 5.30% during the forecast period from 2023 to 2032 (Fig. 2).

The key drivers for the growth of the eco-friendly packaging market include:

- Increased awareness of consumers and their opting for eco-friendly products and packaging solutions;
- Increased pressure from regulatory bodies and government initiatives aimed to reduce plastic waste and promote a circular economy;
- Growing demand for biodegradable and renewable materials such as paper, cardboard, bioplastics, glass and metal;
- Innovation and development of new technologies and designs to enhance the functionality, performance and aesthetics of eco-friendly packaging;

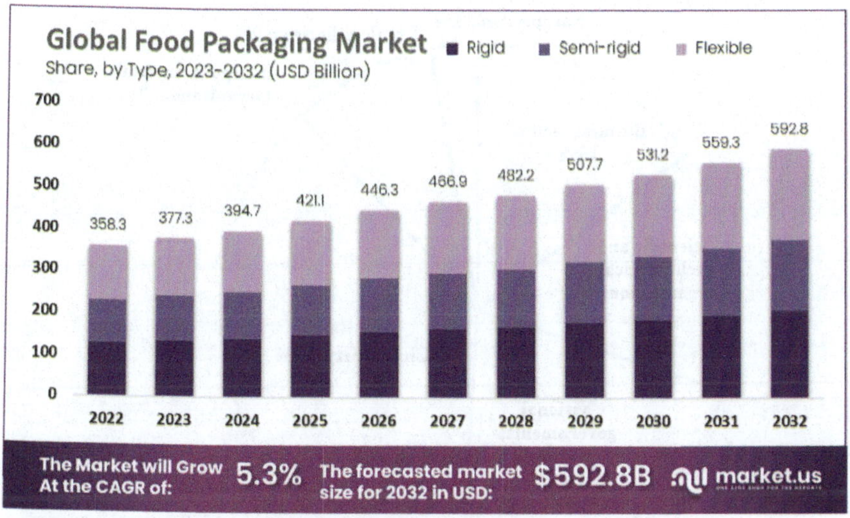

Fig.2. Dynamics of the global food packaging market value

- Cost cut and improved operational efficiency through packaging waste reduction and resource management optimization.

We assume that the state of development of this industry meets the goal of sustainable development, namely, transition to sustainable models of consumption and production.

The symposium held in Oslo in 1994 adopted the notion of sustainable consumption and production as "the use of services and related products that satisfy basic needs and contribute to an improved quality of life, while minimizing the use of natural resources and toxic materials, and emissions of waste and pollutants throughout the life cycle of a service or product so as not to compromise the needs of future generations." [12].

To ensure sustainable consumption and production patterns, sustainable packaging meets the following challenges [13]:

1) sustainable development and efficient management of natural resources;
2) reduction of the waste volume by measures taken to prevent its formation, reduction, recycling and reuse;
3) implementation of sustainable production methods;
4) provision of sustainable public procurement practices in line with national strategies and priorities;
5) assistance to developing countries in increasing their scientific and technological capacity to employ more sustainable consumption and production models;
6) raising awareness among people worldwide about relevant information and knowledge of sustainable development and lifestyles in harmony with nature.

The progress report on the 10-Year Framework of Programmes on Sustainable Consumption and Production Patterns [14] updates the participants of the sustainable consumption and production network, which include governments, civil society, the private sector, and academic institution and international organizations (Fig. 3).

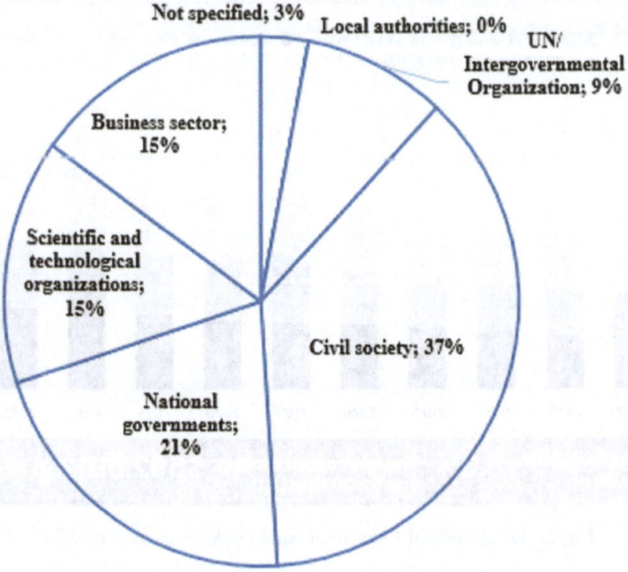

Fig.3. Sustainable production and consumption network

Consider the effect of individual stakeholders on the development of the eco-friendly packaging industry.

Government Regulation

Governments across the globe are taking action to promote eco-friendly packaging and tackle plastic waste. Some of the strategies and initiatives that have been implemented or proposed include:

1) The EU's Circular Economy Action Plan (CEAP): [15] A comprehensive strategy to make all packaging in the EU reusable or recyclable by 2030, reduce the consumption of single-use plastics, and limit the use of certain microplastics.
2) The UK Plastic Packaging Tax: [16] A tax that came into force in April 2022 and applies to plastic packaging manufactured in the UK or imported into the UK with less than 30% recycled plastic. The tax is intended as an incentive for businesses to use recycled material and aims to encourage greater plastic recycling investment.
3) China's Extended Producer Responsibility (EPR) System [17]: A policy framework that requires producers and importers of certain products, such as e-commerce platforms, express delivery companies and beverage companies, to take responsibility for collecting, recycling and disposing their packaging waste.
4) An updated system of extended producer responsibility that came into force in Russia on January 1, 2024, imposes responsibilities for waste management and manufacturing eco-friendly goods on producers and importers. Producers, not the companies that manufacture goods, will be responsible for packaging waste disposal. Under EPR regulations adopted in 2027, producers and importers are obliged to perform 100% recycling of packaging produced. Yet, the rate of packaging waste recycling will increase gradually: 55% in 2025, 75% in 2026, and 100% in 2027 [18].

Civil Society

The report Sustainable Food Packaging in 2040 by UPM Specialty Papers and Smithers consultancy [19] predicts a growing trend in consumer preference for recyclable, compostable or biodegradable packaging. By 2040, experts expect a $20 billion transition from plastic to eco-friendly food packaging [20].

A study by Capgemini consultancy firm on sustainability and variations in consumer behavior revealed that 79% of consumers change their purchasing habits primarily due to concerns about social responsibility, inclusion and environmental effects. Additionally, 53% of consumers and 57% of 18–24 aged consumers have switched to lesser known brands because they are more sustainable. More than 52% of consumers say that they share an emotional connection with companies or organizations they perceive as sustainable [21].

Scientific and Technical Organizations

In 2021, nearly half of all packaging materials were polymer-based, which amounted to 40% share of the global market by value. However, fiber-based packaging is expected to become an alternative to plastics for food packaging due to consumer commitment to sustainable packaging trends.

This trend requires scientific research and innovative solutions.

Business Sector

Businesses aim to adapt to meet the requirements of end consumers, legislative regulation, and respond to current environmental issues, including adoption of modern production technologies for sustainable resource management.

Business sector realizes its objectives through the implementation of sustainable consumption and production projects.

The project for sustainable development in general and its environmental component in particular is a complex change project. The project implemented at an enterprise of the food packaging industry may yield:

- development of a new product (for example, biodegradable packaging), including research (R&D);
- development, acquisition and adoption of a new or improved production technology to significantly reduce adverse environmental effects (the amount of production and consumption waste; air, water and soil pollution; etc.);
- construction of capital facilities (buildings, structures, engineering and transport networks) to ensure green technologies, including renewable energy sources;
- implementation of changes in the structure, processes, personnel of the organization (restructuring projects) in order to improve the production culture.

We consider it reasonable to present the results of sustainable development projects as a classification of the main values with respect to material characteristics (Fig. 4).

Management subjects require special consideration as part of the implementation of projects aimed at achieving one of the goals of sustainable development, that is, sustainable consumption and production in the food packaging industry.

The subjects are as follows:

1) project integration, including:

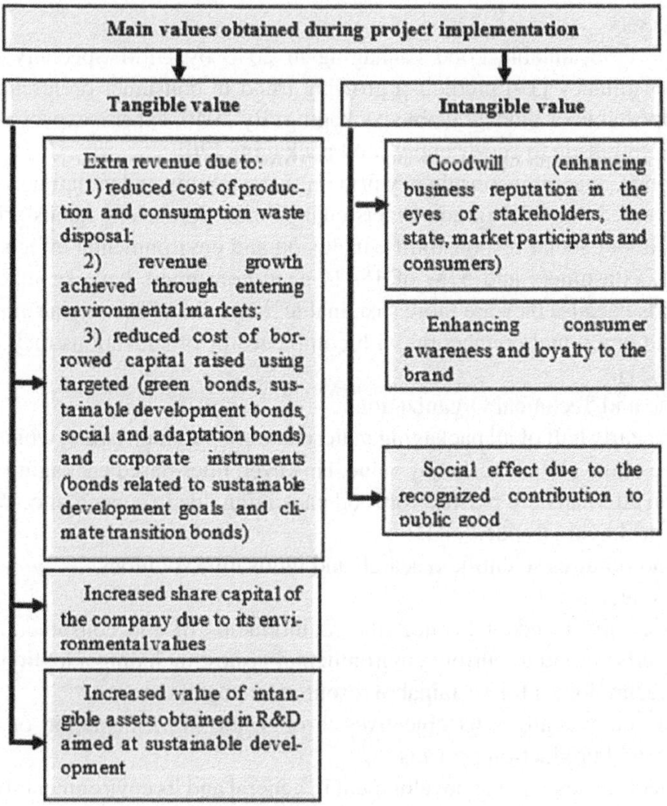

Fig.4. Classification of values yielded by the project

- development and operation of the interaction system for the main project participants (the project initiator, supervisory and regulatory government agencies, financial institutions, educational and research institutions, industry and public associations);
- development and approval of the optimal format and channels of information support to satisfy stakeholders of the project.

2) project content (strategic goals, objectives and processes specified by internal regulations of the enterprise within the framework of the project);
3) project timing (determination of the life cycle and main project phases, development of plans and schedules for project implementation with a sufficient level of project task decomposition and detail);
4) project budget (determination of planned cost estimates, including the structure and amount of project costs for all project stages);
5) project quality (qualification and validation of the main project processes and products – project results);
8) project resources (analysis of the need and availability of material, intangible, financial, and labor resources, including identification of their sources);

7) project risks:

   - internal: generated by project processes and subjects
   - external: generated by external processes and subjects
   - Mitigation and elimination of risks requires their timely identification, analysis, assessment, and management. This approach is confirmed by F.H. Knight in his work Risk, Uncertainty and Profit published in 1921. "The effect of any change which can be foreseen will be adequately discounted in advance, and 'costs' connected with it will be affected in exactly the same way as the corresponding 'values' and no separation between the two will take place" [22].

8) project procurement (planning, competitive procedures, control of supplies and payments).

The above management subjects must be assigned in the matrix of responsibility to the relevant management entities (responsible persons from among the employees and owners of the enterprise).

5 Conclusion

Today's society faces the challenges that require coordinated and systemic action. A resilience agenda should focus on specific. Issues and their interconnectedness. International organizations and national governments set the current agenda, goals and indicators. Educational and research institutions develop technologies and methods of production to solve demands that promote sustainable development. Business implements projects that ensure its growth with regard to the strategic priorities of social development.

Strict time limits necessitate the use of the existing methods and tools that have proven their effectiveness in other fields in order to solve the problems in the context of new challenges. At the same time, the evolution of these methods and practical tools through alternative applications must be constantly advanced with due regard to new theoretical justifications and practical experiments.

References

1. United Nations. The Sustainable Development Goals Report 2023: Special Edition. https://unstats.un.org/sdgs/report/2023/. Accessed 14 Feb 2024
2. The Circularity Gap Reporting. The circularity gap report 2023. https://www.circularity-gap.world/2023. Accessed 14 Feb 2024
3. Deming, E.: Management of New Times: Simple Mechanisms that Lead to Growth, Innovation and Market Dominance. Alpina Publisher, Moscow (2019)
4. Daily Digital Digest 3D News. Renewables covered almost 60% of German electricity demand in 2023. https://3dnews.ru/1098311/vozobnovlyaemie-istochniki-energii-obespechili-597-sprosa-na-elektroenergiyu-v-germanii-v-2023-godu. Accessed 05 Jan 2024
5. Karen Baum. The Business Case for ESG Risk Assessment / Risk Management e-Magazine October 2023. https://www.rmmagazine.com/articles/article/2023/10/09/the-business-case-for-esg-risk-assessment. Accessed 17 Oct 2024
6. European Environment Agency, Reproduction is authorized provided the source is acknowledged. 2023. https://european-union.europa.eu/index_en. Accessed 14 Feb 2024

7. DIN 69901-2:2009 Project management — Project management systems — Part 2: Processes, Process Model. Standartinform, Moscow (2016)
8. United Nations, Report: Science, technology and innovation for sustainable development, 2023. https://unctad.org/system/files/official-document/a78d232_ru.pdf. Accessed 14 Feb 2024
9. Ding, Z., Kovalev, M., Novik, V.: The phenomenon of Economic Development of China. BSU Publishing center, Minsk (2008)
10. Kuleshov, P.: Corporate social responsibility and sustainable development: global experience and concept for Russian enterprises. Bull. MMA **2–2022**, 52–56 (2022)
11. Food packaging market research report 2023. https://market.us/report/food-packaging-market/. Accessed 14 Feb 2024
12. United Nations. Sustainable consumption and production. https://sdgs.un.org/ru/topics/sustainable-consumption-and-production. Accessed 14 Feb 2024
13. United Nations. Goal 12: Ensure sustainable consumption and production patterns. https://www.un.org/sustainabledevelopment/ru/sustainable-consumption-production/. Accessed 14 Feb 2024
14. United Nations: Progress report on the 10-year framework of programmes on sustainable consumption and production patterns. https://www.oneplanetnetwork.org/sites/default/files/russina.pdf. Accessed 14 Feb 2024
15. Circular economy action plan. The European green deal. https://environment.ec.europa.eu/strategy/circular-economy-action-plan_en. Accessed 10 Feb 2024
16. Plastic Packaging Tax. https://www.gov.uk/government/collections/plastic-packaging-tax. Accessed 14 Feb 2024
17. Introduction of Extended Producer Responsibility in China. https://www.assia.nagoya-u.ac.jp/20230122%20ASSIA%20WP22-04%20Introduction%20of%20EPR%20in%20China.pdf. Accessed 10 Feb 2024
18. Interfax. In Russia, in 2024, the environmental tax rate will remaint at the 2023 level. https://www.interfax.ru/russia/939347. Accessed 11 Feb 2024
19. Waste Today Study. Outlines food packaging forecast for 2040. https://www.wastetodaymagazine.com/news/smithers-upm-specialty-papers-2040-sustainable-food-packaging-forecast/. Accessed 11 Feb 2024
20. Smithers. What will sustainable food packaging trends look like in 2040?. https://www.smithers.com/resources/2022/may/sustainable-food-packaging-trends-2040-upm, Accessed 11 Feb 2024
21. The Report Covers Global Sustainable Packaging Industry Trends and Companies. https://www.mordorintelligence.com/industry-reports/sustainable-packaging-market, Accessed 07 Feb 2024
22. Knight, F.: Risk, Uncertainty and Profit. Delo, Moscow (2003)

HR Ecosystem of the Fishing Industry: From Structural to Workforce Complexity

Oksana Ogiy(✉)

Kaliningrad State Technical University, Kaliningrad, Russia
oksana.ogij@klgtu.ru

Abstract. The paper studies industry human resource management based on the ecosystem approach. Using the Russian fishery complex study materials, two fundamental factors of its HR ecosystem development were identified: the structural complexity of the industry and the workforce complexity. Expert assessment shows that the problems of interaction and consistency have been identified between the main structural elements that form the industry ecosystem. The results of a sociological survey of workers in the fishing industry (N-1005) served as empirical basis for developing an approach to assessing the complexity of the workforce. This approach allows to analyze not only the abilities and personal characteristics of employees, but also to assess the prospects for their activity. Employee qualities indicators significance and interdependence were determined using the Weight of Evidence (WoE) method. These results allow us to include the assessment of the workforce as a diverse and complex ecosystem entity into HRM practice. The level of coordination among industry participants and the workforce complexity have a significant impact on the ability to implement an industry development strategy. The best methodology is HRM based on an ecosystem approach given the complexity of the research and its management practice.

Keywords: HR-Ecosystem · Fishing Industry · Structural complexity · Workforce · Capabilities · Values · Motivation

1 Development of HR Ecosystems: Industry Perspective

Researchers in various aspects of business are attracted by ecosystem approach today; the growing popularity of which is understandable. An ecosystem is a methodological construct that allows to comprehensively consider multidimensional environments. This valuable property is especially relevant when studying the problems of managing transnational and extraterritorial structures. The complexity of managing such structures is determined by a number of factors: blurred organizational boundaries, the need to harmonize various management practices, differentiation of human capital, cultural differences [1–4].

In a review of research on HR ecosystems, Snell and others identified three emerging trends in the field: (1) technological intermediation, (2) volatility and adaptation, and (3) the challenge of shared governance [5]. Problems of adaptation, coordination and joint

R. Polyakov (Ed.): EcoSyst 2023, LNNS 1092, pp. 175–193, 2024.
https://doi.org/10.1007/978-3-031-67354-2_20

management in relation to knowledge-intensive organizations have been identified [6]. Researchers found that the main contradictions are formed between four subsystems: strategy, capabilities, composition and culture. Moreover, these contradictions manifest themselves both within and between levels of the economy; for example, within and between organizations at the national level [6].

In paper [7], the main problems in the development of HR ecosystems are defined as the complexity of organizational structures and the coordination of different employment models.

Meanwhile, researchers almost unanimously note that the ecosystem approach makes it possible to establish processes of mutual adjustment and compliance with changing strategic needs [5, 8–10]. Along with that, the use of artificial intelligence technologies significantly enhances the analytical and management advantages of the ecosystem approach [11–13].

The sectoral aspect is extremely important in economic development problems consideration. An industry is an object that integrates at least three of its aspects: the economics of firms and corporations; regional economy; market of products and technologies. It means that the degree of consistency/fragmentation of the industry's economy determines the development opportunities at the micro and macro levels of the economy. Research shows that within the same corporation or group of enterprises one can observe organizations and teams that are at different development levels and have different development rates. The importance of the systemic evolvement of mesoeconomics systems in the modern economy is comprehensively and deeply considered in the works of G. Kleiner [14, 15]. Based on his theory of mesoeconomics systems, we can conclude that the unsystematic nature of the sectoral economy feeds the fragmentation of the society of the regions of presence, the professional community, and therefore a certain part of the national economy and macro society.

Number of interconnected enterprises and organizations are the subjects of industry HR ecosystem study; the behavior of these enterprises is considered simultaneously as a group of objects and a group object. In this context, an important role belongs to the study of factors: 1) the structural complexity of the industry ecosystem and 2) the diversity of its human resources, expressed in the abilities and behavior of workers. Both factors form the strategic potential of the industry and deserve special comprehensive study.

2 Structural Complexity of Russian Commercial Fisheries

Modern industrial fishing is a technologically intensive industry. High technology is determined by the requirements of high productivity, efficiency and waste-free production. This predetermines the presence of a highly productive fishing fleet; technologically advanced reproduction and processing facilities with a high degree of automation; modern port, refrigeration, warehouse infrastructure; productive management systems for fishing and processing. In addition, fishery activities are knowledge-intensive since they directly depend on scientific forecasts of fishing prospects, assessment and methodology for managing stocks of aquatic biological resources, as well as the development of effective measures for their reproduction.

In the post-Soviet period, the Russian fishing industry developed according to the resource type, focusing on the extraction of aquatic biological resources with an extremely low share of deep processing. The fishing fleet, processing equipment and technologies, and port infrastructure are physically outdated and went out of style. Sectoral economy low level of diversification, the dominance of the raw material strategy, weak participation in global value chains and technological stagnation have created limited demand for human capital; and as it turns not in the quantitative aspect (labor), but in the qualitative aspect (knowledge, skills, competence, involvement).

Institutional changes taking place over the past two decades in the fishery complex (updating fisheries legislation; the policy of "keel quotas", "quotas for industrial processing capacities"; expansion and intensification of state support measures) stimulated the renewal of the fishing fleet, the commissioning of new coastal capacities processing plants, refrigeration complexes. Commercial aquaculture is also intensively developing-production volume has increased 4 times over the past 8 years.

These changes led to an increase in technological complexity and diversification of fishing activities, and therefore innovation demand began to develop; the policy of "technological sovereignty" further strengthened these processes.

Such a scenario for the development of an industrial economy places an increased demand on available human resources, which will increase in the coming decades. In this regard, a number of fundamental questions arise for the industry HR ecosystem:

the nature of the sectoral structure, the level and intensity of relations between its elements;

whether they are transformed into high productivity/effectiveness;

how structural complexity and workforce complexity correspond to each other and how to ensure balance;

if the knowledge and skills accumulated by the industry sufficient to solve the problems of "breakthrough" development;

what is HRM role in reconciling these aspects of complexity and implementing global industry strategy.

The structural organization of Russian industrial fishing is determined by multidimensional space; it includes several group management subjects. They are conventionally called "subsystems" of the ecosystem.

The first subsystem is represented by the Federal Agency for Fishery central apparatus and its territorial departments, formed according to the "basin" principle. According to this principle each of the eighteen territorial departments oversees certain parts of the country's eight fishery basins. The fishery basin includes seas and lakes with the basins of rivers flowing into them, as well as other water bodies of fishery importance. This cluster represents the public sector and the federal level of regulation.

The second subsystem of the industry ecosystem consists of government bodies of the constituent entities of the Russian Federation. They are represented by relevant agencies or ministries within regional governments and administrations that promote the development of fisheries in their subordinate territory. This subsystem also represents the public sector, but has significantly fewer regulatory functions. It is designed to promote fisheries activities within regional boundaries and ensure their integration into the regional development strategy.

The third subsystem is formed by enterprises and organizations of various ownership forms, which make the production, product, and commodity architecture of the fisheries market. This is the most numerous and diverse group entity in the ecosystem. It includes ship-owners of fishing vessels, mining and processing enterprises, port and warehouse infrastructure organizations, fish farming enterprises, as well as their alliances (associations and unions). Industry public organizations are very numerous and diverse, but there are very large players among them who have a significant impact on the regulatory sector of the ecosystem. For example, the All-Russian Association of Fishery Enterprises, Entrepreneurs and Exporters unites organizations that account for about 90% of the fish catch in Russian territorial waters.

Within the framework of our study, the subsystem of enterprises and organizations is important as it represents the primary level at which HRM practices develop and the complexity of the workforce is formed.

The fourth subsystem is a complex of scientific and educational organizations. In Russia it is represented by specialized industry organizations, the founder of which is the Federal Agency for Fishery (five universities and a large territorially diversified research organization - the Russian Federal Research Institute of Fisheries and Oceanography), as well as institutes and research teams and departments of other universities and research institutes dealing with individual problems of fisheries and fish farming. This subsystem plays an important role in at least two aspects: 1) the development of future abilities and competencies of industrial fishing; 2) development and transfer of industry innovations. In the development of any social ecosystem, an educational research cluster should play the role of a system-forming factor.

In the study of industry structural complexity, it is important to assess the level of interaction within and between each of the identified subsystems.

The coherence and intensity of interaction between organizations within each group entity reflect its integrative ability and the degree of internal coherence. Internal consistency is important to maintain the pace of development and technological innovation. Integrative ability determines the significance of the "collective player" in the ecosystem and the degree of possible contribution to the implementation of the industry strategy.

The relationships between group entities, the quality and level of their interaction determine the industry ecosystem contribution to national development, and sometimes to the global economy. At this level, coherence turns into synergistic development.

To study the nature of intra- and intergroup relations, paper utilizes the results of 37 experts survey. They were asked to evaluate seven types of relationships between subjects (competition, co-production, cooperation, contacting, coordination, co-evolution, consolidation) on a scale from 0 - "no relationship", to 1 - "the maximum possible expression of the relationship". The assessment was carried out taking into account two factors: the presence of documentary and other material evidence of the relationship (product, formal organization, contract, road map, etc.), as well as the intensity of the relationship, expressed in their frequency and regularity. Assessment subject was a set of organizations interactions within and between the subsystems. The results were compiled into a matrix (Table 1). The indicator of intragroup relations of the subsystem that develops between its constituent elements is highlighted in bold italics.

The results presented in the relationship matrix allow us to draw a number of conclusions.

Table 1. Matrix of Expression of Intragroup and Intergroup Relations in the Fishery Ecosystem

Types of Relationships	1st subsystem	2nd subsystem	3rd subsystem	4th subsystem
Competition				
1st subsystem	***0.3***	0.4	0.03	0.01
2nd subsystem	0.4	***0.6***	0.04	0.02
3rd subsystem	0.03	0.04	***0.8***	0.1
4th subsystem	0.01	0.02	0.1	***0.6***
Contracting				
1st subsystem	***0.7***	0.6	0.5	0.6
2nd subsystem	0.6	***0.5***	0.6	0.5
3rd subsystem	0.5	0.6	***0.7***	0.5
4th subsystem	0.6	0.5	0.5	***0.5***
Co-production				
1st subsystem	***0.8***	0.8	0.4	0.5
2nd subsystem	0.8	***0.2***	0.6	0.5
3rd subsystem	0.4	0.6	***0.3***	0.4
4th subsystem	0.5	0.5	0.4	***0.5***
Coordination				
1st subsystem	***1***	0.6	0.4	0.3
2nd subsystem	0.6	***0.4***	0.3	0.4
3rd subsystem	0.4	0.3	***0.3***	0.2
4th subsystem	0.3	0.4	0.2	***0.1***
Cooperation				
1st subsystem	***0.9***	0.5	0.4	0.4
2nd subsystem	0.5	***0.2***	0.4	0.4
3rd subsystem	0.4	0.4	***0.2***	0.2
4th subsystem	0.4	0.4	0.2	***0.3***
Coevolution				
1st subsystem	***0.6***	0.4	0.3	0.2
2nd subsystem	0.4	***0.3***	0.4	0.3
3rd subsystem	0.3	0.4	***0.1***	0.2
4th subsystem	0.2	0.3	0.2	***0.2***
Consolidation				
1st subsystem	***0.6***	0.3	0.1	0.1

(continued)

Table 1. *(continued)*

Types of Relationships	1st subsystem	2nd subsystem	3rd subsystem	4th subsystem
2nd subsystem	0.3	***0.1***	0.2	0.2
3rd subsystem	0.1	0.2	***0.05***	0.01
4th subsystem	0.1	0.2	0.01	***0.05***

1. When relationships are not associated with direct subordination, as, for example, in subsystem 1, the interactions development level is quite weak even within subsystems.
2. On average competitive relations are more manifested than cooperative relations in terms of territorial and organizational structuring (subsystems 2 and 3), as well as in education and science (subsystem 4). This is not a rational strategy for the taken industry complex. There are studies [16] that indicate that economic actors with an increased level of internal competition are less able to withstand adverse external influences and are less effective.
3. As the level of relations consistency increases (from the lowest (contracting) to the highest (consolidation), they occur less and less often. This is manifested both in the intragroup circuit and in the relations between subsystems. This state of affairs indicates a low level of integration processes in the industry ecosystem.
4. Against the background of rather high structural complexity, the development and diversity of relationships is at a rather low level. The nature of the structure and relationships in the industry allows us to characterize it as a "protocluster". Since the relationships between subsystems develop in the form of information interaction and technological connections, which are unstable and short-term in nature.

Thus, the fishery ecosystem is at the initial stage of its development. In this regard, it is important to answer a number of questions. How does high complexity due to competing priorities within subsystems and weak integration processes between them affect HRM? Are these disadvantages offset by the complexity of the workforce?

3 Workforce Complexity

The Russian fishery complex is characterized by rapid transformations, driven by high competition and an ambitious strategy in the innovation field. The success of implementing such a strategy depends on the active and passive transformation behavior of economic entities [17].

The ability to achieve goals and implement strategies is based on the idea of the best "desired" state of the organization. Therefore, the idea of effectiveness has a value-based nature, i.e. depend on our opinion about what is proper and important. Many studies point to the connection between performance and values. Thus, "pathologies of goal setting" are described as a result of the alienation of goals from values [18]; a close connection has been established between systemic economic cycles and the progressive change of value memes in society [19]; changes in the nature of the organizational environment under the influence of participants' value systems in economic activity are described in the paper [20].

The ability to achieve goals is an equally important "behavioral characteristic" of an organization's performance. This ability is manifested in decisions made and actions taken. They are as meaningfully dependent on values as goals. But their intensity and direction are determined by motivation [21]. The normative value structure of an organization is also closely related to the phenomenon of motivation, and is considered by researchers as an integral structure [22, 23].

Thus, the study for the industry development prospects is inevitably concerned with the way the complex set of abilities, values and motivational attitudes of different employees groups correspond to and support it. In this context, the HR ecosystem focuses not on an impersonal workforce, but on an ecosystem of people working collaboratively.

For example, in [24], based on the "ability-motivation-opportunity" analysis, it was proven that initiatives to increase HR motivation build up motivation of line managers; expanding HR capabilities improves the perceived abilities, motivation and capabilities of line managers, and indirectly the effectiveness of implemented HR technologies. The opportunities provided by the management system to line managers become a decisive factor in increasing the efficiency of the implementation of personnel management technologies. A study of the phenomenon of transformational leadership [25] showed that significant factors influencing motivation are the vision and authority of employees, i.e. factors related to HRM.

Experimental data on value orientations, value profiles, types of motivation and abilities of workers in various subsystems of the fishery complex were obtained using the software "System of Management of Labor Potential of the Fishery Complex" [26] with a module for surveying workers using 134 indicators. The total number of workers surveyed was 1005 people aged from 17 to 70 years: 331 women and 674 men; 233 students of specialized universities and colleges, 362 workers of fishing enterprises and institutions, 410 crew members of fishing vessels. As a result, a database was obtained with each respondent's 134 indicators assessment, which form 47 indicators. The composition of the assessed factors is presented in detail in paper [27]. The survey covered representatives of the industrial fisheries of the five largest fishing basins in Russia. A study of a number of properties has been carried out using known methodologies. Personal attributes have been examined using the Big Five Inventory-2 [28]; values - the Sh. Schwartz questionnaire [29]; type of motivation - by the Motype test [30]. The remaining parameters have been calculated based on the answers of employees to 54 questions of a specialized questionnaire.

The survey covered representatives of the industrial fisheries of the five largest fishing basins in Russia. Automated data processing and analysis were carried out in Loginom Community and SPSS Statistics.

Weight of Evidence methods [31] and the calculation of Information Value were used for comprehensive assessment of the fishery complex workforce and establish dependencies between the characteristics of workers and their activity.

Coefficient WoE was used to form final classes, where each observation contains a set of features. Next, the range of values of each variable is quantized into specified intervals (for example, age groups in increments of 5 or 10 years; job levels; test scoring intervals, etc.) Each observation of a particular employee characteristic was matched binary output variable: "Positive cases" (P) or "Negative cases" (N). The endowment of a certain characteristic of an employee (group of employees) with a P or N characteristic occurs depending on the significance of this characteristic for a specific aspect of performance (productivity, quality, degree of task completion, degree of strategy implementation). P_i and N_i value is calculated for each interval.

Based on dividing the entire range of changes in a particular employee characteristic into several initial classes, the Coefficient WoE is calculated for each of them:

$$WoE_i = \ln \frac{F-}{F+}, \tag{1}$$

where i – primary class index; F^- - relative frequency of occurrence of negative cases (N_i/N); F^+- relative frequency of occurrence of positive cases (P_i/P) in the class.

If the relative proportion of negative outcomes (F^-) is greater than positive ones (F^+), then the value under the logarithm sign > 1 and WoE has a positive value, otherwise it is negative. Hence, WoE < 0, indicates a greater likelihood of a positive outcome, and WoE > 0 - negative.

To calculate the WoE coefficient in order to determine the probability of manifestation of one or another type of motivation in different work differentiation conditions, observations were selected where a specific type of motivation is dominant. This means that all test results with a value above the third quartile of the distribution were coded as the output variable Positive cases ($x > Q_3$). Next, the dependence of the variables "dominant type of motivation" on other independent variables characterizing the surveyed workers was sequentially checked. Selected results of the WoE analysis are presented below.

The manifestation of motivation types in different age groups of workers is presented in Table 2.

According to these data, the most age-dependent are Occupational, Proprietor and Avoidance motivations. Obviously in the first case, the older the employee, the more likely the manifestation of Occupational motivation. Proprietor motivation manifests itself in middle-aged workers from 30 to 50 years old. Avoidance motivation which is "undesirable" motivation for the company is most likely to be found among workers under 30 years of age and over 60 years of age.

We tested the assumption about the connection between the complexity and variability of job responsibilities and the employee's motivational attitudes by calculating WoE for each type of motivation for intervals formed by the distribution of workers across three levels of jobs they occupy: senior-level, midlevel, low-level. The results are presented in Fig. 1.

The first thing that focuses attention is the clear manifestation of the typicality of motivation among employees of higher and lower levels. And a weak manifestation of the dominance of one motivation or another in middle positions.

For leaders and top managers the dominant types of motivation are: Patriotic (WoE = –1.03), Occupational (WoE = –0.76), Proprietor (WoE = –0.5). Among workers and maintenance personnel, dominance is typical by Instrumental (WoE = –0.9) and Avoidance (WoE = –0.99) motivation types and very unlikely manifestation of Occupational (WoE = 1.91) and Patriotic (WoE = 2.13) types.

Table 2. The results of calculating the WoE coefficient of different ages employees motivation types.

Age Brackets	Occupational	Instrumental	Patriotic	Proprietor	Avoidance
Age < 20	0.177	−0.023	0.005	0.466	−0.222
20 ≤ … < 25	0.308	−0.669	0.488	0.403	−0.130
25 ≤ … < 30	0.607	0.318	0.320	1.974	−0.158
30 ≤ … < 35	0.316	0.580	−0.326	−0.471	0.123
35 ≤ … < 40	0.193	0.054	−0.242	−0.096	0.567
40 ≤ … < 45	−0.312	0.735	−0.269	−0.432	0.138
45 ≤ … < 50	−0.807	−0.149	−0.530	−0.309	0.247
50 ≤ … < 55	−1.114	−0.277	0.218	1.167	0.418
55 ≤ … < 60	−0.139	1.281	0.524	0.744	0.555
60 ≤ … < 65	−0.609	4.199	−0.895	1.096	−0.275
Age ≥ 65	−0.426	3.862	−0.223	0.744	−0.093

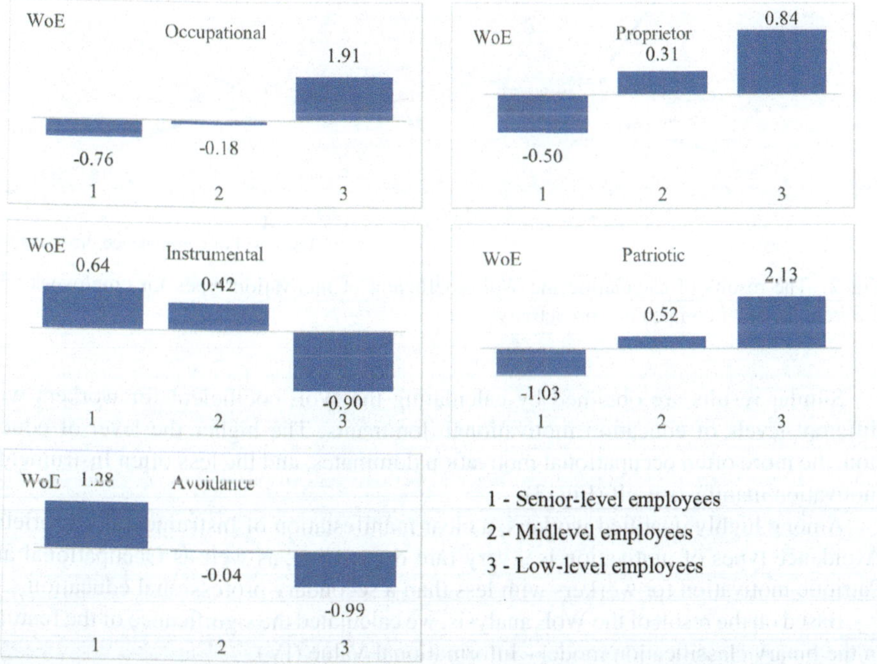

Fig. 1. The results of the calculation of the WoE coefficient by motivation types for different position levels.

Thus, we can talk about the connection between the motivational attitudes of workers and the positions they occupy at the top and bottom of the job hierarchy.

We will also present separate results of testing the dominance of motivation depending on the employee's competence indicators.In the first case, this is competence-based proactiveness (Fig. 2), in the second, the level of education (Fig. 3) of the employee.

Grouping assessments of competence-based proactivity from minimum (min 0–0.2) to maximum (max 0.81–1) and calculating the WoE coefficient for the dominant types of motivation lead to several conclusions. The higher the level of competence-based proactivity of an employee, the more likely Occupational and Patriotic motivations will be manifested. This is clearly visible in the graph (Fig. 2). The lower the employee's competence-based proactivity, the more likely the dominance of Instrumental motivation (WoE = −0.9).

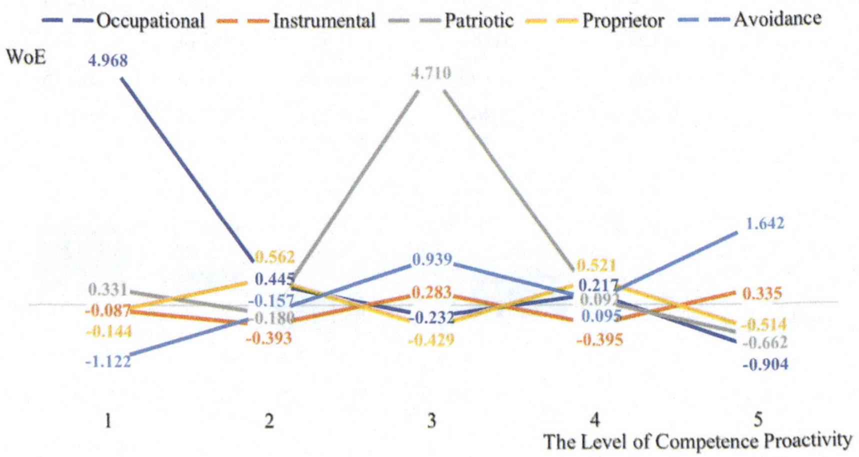

Fig. 2. The results of calculating the WoE coefficient of motivation types for employees with different levels of competence proactivity.

Similar results are obtained by calculating the WoE coefficient for workers with different levels of education motivational dominants. The higher the level of education, the more often occupational motivation dominates, and the less often Instrumental motivation manifests itself (Fig. 3).

Among highly qualified workers, a clear manifestation of Instrumental, Proprietor, Avoidance types of motivation is a very rare occurrence, as well as Occupational and Patriotic motivation for workers with less than a secondary professional education.

Based on the result of the WoE analysis, we calculated the significance of the features in the binary classification model - Informational Value (IV).

$$IV = \sum_{i=1}^{k} \left\{ \left(\frac{N_i}{N} - \frac{P_i}{P} \right) \times WoE_i \right\} \quad (2)$$

Moreover, the significance of the sign if: IV < 0.02—absent; 0.02 ≤ IV < 0.1—low; 0.1 ≤ IV < 0.3—middle; IV ≥ 0.3—high.

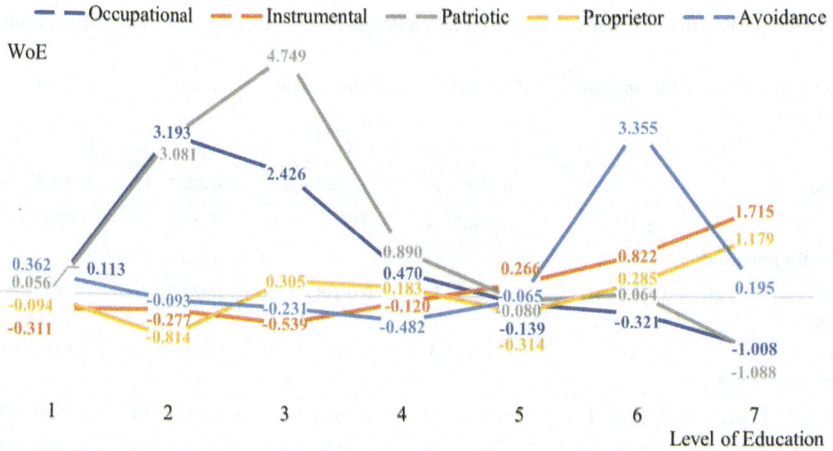

Fig. 3. The results of calculating the WoE coefficient of motivation types for employees with different levels of education. *1 - secondary general and secondary education (school), 2 - vocational courses and schools (without secondary education), 3 - primary vocational education, 4 - secondary vocational education, 5 - higher vocational education, 6 – university level, residency, 7 - higher qualification (academic degree)*

This method of determining the significance of an employee's characteristics for certain performance criteria seems to be quite optimal and representative, since it maximizes the significance of the attribute in the binary model and maximizes the uniformity of filling the intervals. Next, we present the results of calculating the information value of differentiated assessment of employees in two areas.

Firstly, to demonstrate one or another type of motivation (Table 3).

Secondly, for workers with dominance of certain value systems. Table 4 shows the results of determining the information value of various characteristics of employees to form a value profile, which we tentatively called "Effective Leader".

It should be noted that we can take any analyzed value profiles, depending on the purpose of their study. To identify employees with the "Effective Leader" profile, we used the results of a survey of the value profile of employees using the Schwartz questionnaire. From all observations, we selected those employees whose value profile is dominated ($x > Q_3$) by three individual priorities: Achievement, Self-Direction, Benevolence. Then we calculated the WoE coefficient of the interval assessment of each characteristic of employees with the "Effective Leader" value profile. The results of calculating information significance are summarized in Table 4.

As is the case with different types of motivation (Table 3), and certain value systems (Table 4), individual indicators from different groups have a high value.

Relationships and mutual influence of employee characteristics is not analyzed and generalized in this article. The point is to demonstrate that deep differentiated assessment of personnel is possible and very informative in itself. But in the context of the work, it seems to be more interesting how the identified complexity of the workforce can be integrated with the structural aspects and strategies of organizations to improve HRM and develop HR ecosystems.

Table 3. Results of Calculation of Employee Properties IV for Types of Labor Motivation.

Employee Indicator	Occupational		Proprietor		Instrumental		Patriotic		Avoidance	
	(IV)	*	(IV)	*	(IV)	*	(IV)	*	(IV)	*
Age	0.2258	M	**0.8791**	B	0.2244	M	**0.4307**	**H**	**0.3463**	**H**
Gender	0.0215	L	0.0002	A	0.0002	A	0.0050	A	0.0102	A
Health potential										
Current physical state	0.0287	L	0.0653	L	0.0038	A	0.0005	A	0.0153	A
Health self-assessment	0.0259	L	0.0134	A	0.0481	L	0.2074	M	0.1249	M
Chronic diseases	0.0559	L	0.2549	M	0.1050	M	0.1171	M	0.1860	M
Health risk factors	0.2430	M	0.2156	M	0.2255	M	0.2269	M	0.2312	M
Physical activity and exercise	0.0119	A	0.0551	M	0.0509	L	0.0649	L	0.1619	M
Personality traits										
Subjective assessment of the happiness level	0.0745	L	0.2134	M	0.1303	M	0.0781	L	0.1571	M
Personality Profile (BFI-2):										
Extraversion	**0.3956**	**H**	**0.6449**	**H**	0.2739	M	0.0861	L	**0.6779**	**H**
Agreeableness	0.1902	M	**0.3127**	**H**	**0.3397**	**H**	**0.5687**	**H**	0.1759	M
Conscientiousness	0.2238	M	0.1282	M	0.2024	M	**1.0435**	**H**	0.0572	L
Neuroticism	0.1715	M	0.1761	M	**0.3532**	**H**	0.2348	M	**1.5568**	**H**
Openness to Experience	**0.3964**	**H**	0.2899	M	0.1277	M	0.2479	M	**1.8868**	**H**
Expertise										
Skill level	0.1079	M	0.2053	M	0.1069	M	0.1014	M	0.2478	M
Competence proactivity	**0.5480**	**H**	**0.3481**	**H**	**0.3434**	**H**	**0.3085**	**H**	**0.6576**	**H**
Length of service and experience	**0.3298**	**H**	**1.1455**	**H**	0.2021	M	**0.9574**	**H**	0.2960	M
Field of employment in the industry	0.0702	L	**0.4206**	**H**	0.0712	L	0.0343	L	0.2524	M
Position level	0.0240	L	0.2298	M	0.1244	M	0.2888	M	**0.4153**	**H**
Value Profile										
Security	0.1996	M	**0.8009**	**H**	0.1082	M	0.1254	M	0.1702	M
Tradition	0.0363	L	0.2923	M	0.0758	L	0.2163	M	**0.3737**	**H**
Self-Direction	**0.3402**	**H**	0.1009	M	0.0339	L	0.0709	L	**0.4177**	**H**

(continued)

Table 3. *(continued)*

Employee Indicator	Occupational		Proprietor		Instrumental		Patriotic		Avoidance	
	(IV)	*	(IV)	*	(IV)	*	(IV)	*	(IV)	*
Achievement	**0.5630**	**H**	**1.1804**	**H**	0.1886	M	0.2381	M	0.1708	M
Hedonism	0.1700	M	0.1623	M	**0.5006**	**H**	**0.3085**	**H**	0.1542	M
Benevolence	0.0525	L	0.1923	M	0.1903	M	0.1543	M	0.1534	M
Conformity	0.0751	L	**0.5084**	**H**	0.1004	M	**0.3770**	**H**	0.1661	M
Universalism	0.1141	M	0.1196	M	**0.5902**	**H**	0.0516	L	0.0743	L
Stimulation	0.1128	M	**0.3362**	**H**	0.1580	M	0.1366	M	0.1123	M
Power	0.2008	M	**0.4089**	**H**	0.1044	M	**0.4143**	**H**	0.0766	L
% of Positive cases	13.86		7.64		18.39		9.72		4.66	
% of Negative cases	86.14		92.36		81.61		90.28		95.34	

* factor significance levels: H-High, M-Middle, L-Low, A-Absent

Table 4. The Importance of Employee Characteristics for the Formation of a "Effective Leader" Value Profile.

Employee Indicator	Informational Value (IV)	Indicator significance
Age	0.2046	Middle
Gender	0.0066	Absent
Health potential		
Current physical state	0.0023	Absent
Health self-assessment	0.0384	Low
Chronic diseases	0.1206	Middle
Health risk factors	**0.3217**	**High**
Physical activity and exercise	0.0404	Low
Personality traits		
Subjective assessment of the happiness level	**0.3499**	**High**
Personality Profile (BFI-2):		
Extraversion	**0.5907**	**High**
Agreeableness	0.1971	Middle
Conscientiousness	0.2253	Middle
Neuroticism	**0.4321**	**High**

(continued)

Table 4. *(continued)*

Employee Indicator	Informational Value (IV)	Indicator significance
Openness to Experience	0.2506	Middle
Expertise		
Skill level	0.2018	Middle
Competence proactivity	**0.3716**	**High**
Length of service and experience	0.2113	Middle
Field of employment in the industry	0.2248	Middle
Position level	0.2398	Middle
Type of work motivation		
Occupational	**0.4244**	**High**
Instrumental	0.1095	Middle
Patriotic	**0.5558**	**High**
Proprietor	**1.0265**	**High**
Avoidance	**0.6413**	**High**
Number/% of Positive cases	108	10.75
Number/% of Negative cases	897	89.25

4 Complexity and Coherence as the Basis for the Development of the HR Ecosystem

All organizations of the fishery complex, whose performance indicators were assessed by experts and whose employees were surveyed, were distributed according to the degree of development of HPW practices and the level of technology used and inter-organizational cooperation. The assessment was carried out according to two groups of criteria. The first group characterizes the internal strategy (IS) and describes the "HR architecture" [32] of the organization and includes the criteria: IS1 - the amount of funding by the organization for personnel development programs, IS2 - variability of remuneration systems, IS3 - variability of employment methods and work regimes, IS4 - the use of human resources in inter-organizational networks. The criteria of the second group characterize the external strategy (ES) and describe the behavior of the organization in the industry and characterize the level of its technology and strategy at the network level. These include: ES1 - level of production technology and automation, ES2 - dependence of profitability on the industry; ES3 - density of interorganizational connections, ES4 - strategic activity at the network level.

The fact that is noteworthy is a very small proportion of industry organizations whose cumulative assessment value according to the criteria of the "HR architecture" and "Conduct in the industry" groups is $\geq$ 60 percentile of the distribution (Fig. 4).

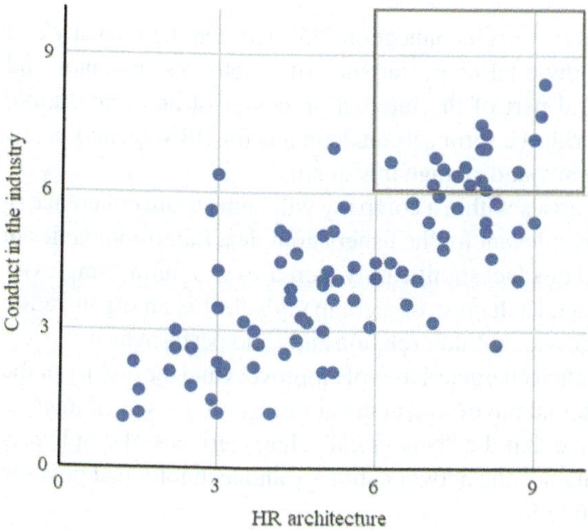

Fig. 4. Distribution of fishery organizations.

There is also a clearly expressed higher level of development of internal management practices in comparison with the activities of organizations in the external environment. Thus, there are 1.8 times more organizations with a high level of development of HR architecture than with active industry behavior. The share of organizations with a high level of development of internal (IS) and external (ES) strategies is very insignificant and amounts to 19, 5%. Next, we made an attempt to determine the degree of connection between the management practices of fishery organizations and certain dominant characteristics of the workforce (W1 -Age, W2 - Health risk factors, W3 - Employees with the "Effective Leader" value profile, W4 - Occupational motivation, W5 - Patriotic motivation, W6 - Competence proactivity, W7 - Skill level) at the top and middle levels of jobs. The exclusion of workers and service personnel from the sample was made to increase the concentration of strategically significant jobs in the sample. To study the relationship, we used the following statistics: Conjugacy Coefficient (C), Cramer's V (V), Uncertainty coefficient (U). The Uncertainty coefficient is given in Tables 5 and 6 as a symmetric measure.

Table 5. The relationship between the quality of the workforce and the HR architecture of fishery organizations.

		IS1			IS2			IS3			IS4	
	C	*V*	*U*	*C*	*V*	*U*	*C*	*V*	*U*	*C*	*V*	*U*
W1	0.444	0.248	0.060	0.693	0.279	0.125	0.454	0.294	0.139	0.764	0.482	0.385
W2	0.830	0.278	0.274	0.396*	0.249*	0.187	0.427	0.272	0.042	0.426	0.396	0.127
W3	0.692	0.563	0.473	0.421	0.589	0.117	0.533	0.363	0.233	0.617	0.784	0.180
W4	0.526	0.309	0.126	0.437	0.485	0.073	0.600	0.433	0.127	0.595	0.279	0.169
W5	0.593	0.301	0.092	0.540	0.262	0.116	0.508	0.278	0.171	0.559	0.275	0.155
W6	0.685	0.746	0.581	0.612	0.253	0.227	0.634	0.310	0.227	0.787	0.619	0.453
W7	0.865	0.460	0.367	0.568	0.334	0.276	0.409*	0.317*	0.108	0.662	0.468	0.455

Note. When $p < 0.01$ table data was not marked, when $p < 0.05$ marked *.

The obtained data suggest that there is a connection [33] between the intensity and variety of HRM practices and individual characteristics of employees in senior and middle positions. Being an integral part of the internal processes of an organization, the quality and diversity of the workforce form certain demands for HRM technologies, which, in turn, influence this diversity and change this quality.

Exploring the complexity that arises within a company will help ensure coherence in business strategy which seems very relevant for the fishery complex. Large companies in the industry have fishing fleets and this fact significantly increases structural complexity and coherence requirements; in fact, each crew of a company's fleet is an organization with its own structure, goals, processes, culture, relationships and performance.

The connection between individual characteristics of employees and the quality of the organization's activities in the external environment cannot be as obvious as in internal processes. At the same time, we see that the "behavioral" characteristics of employees in senior and middle positions influence the network strategy and technological position in the industry, and vice versa (Table 6).

Table 6. Connection between labor force quality and industry behavior of fishery organizations.

	ES1			ES2			ES3			ES4		
	C	*V*	*U*	*C*	*V*	*U*	*C*	*V*	*U*	*C*	*V*	*U*
W1	0.283	0.295	0.075	0.242	0.176	0.073	0.423	0.330	0.043	0.185	0.188	0.046
W2	0.209	0.151	0.039	0.172*	0.123*	0.028	0.129	0.130	0.012	0.268	0.161	0.037
W3	0.771	0.365	0.227	0.291	0.305	0.077	0.553	0.652	0.451	0.770	0.698	0.225
W4	0.543	0.647	0.397	0.333	0.204	0.155	0.409	0.449	0.128	0.698	0.563	0.245
W5	0.436	0.280	0.097	0.400	0.308	0.123	0.315	0.192	0.041	0.474	0.269	0.113
W6	0.862	0.473	0.274	0.660	0.359	0.159	0.520	0.430	0.232	0.486	0.394	0.292
W7	0.618	0.454	0.194	0.247	0.370	0.116	0.202*	0.206*	0.057	0.461	0.367	0.155

Note. When $p < 0.01$ table data was not marked, when $p < 0.05$ marked *.

Using the complexity of the workforce as a management methodological basis offers good prospects for bringing SHRM into line with the environment.

Research into the substantive aspect of the interaction between the complexity of the workforce and the internal and external strategies of the organization still requires a lot of effort. At the same time, we can already be guided by the fact that the complexity of management should reflect the complexity of the workforce and its consistency with the organization's strategies.

The development of an HR ecosystem can help solve problems of coherence and intensification of fisheries development, compensate for structural complexity and help to improve the strategies coherence.

5 Conclusions

Studying processes in industry complexes opens up significant prospects for improving HR ecosystems.

An industry is interacting actors complex object, combining the processes of differentiation and integration. This will help to understand the role of HRM better in the context of the complexity of defining the boundaries of organizations and managing them beyond their boundaries.

The changing and dynamic industry environment places demands on management for greater flexibility, innovation, differentiation and coherence. In this context, the HR ecosystem becomes a development key driver and the basis of strategic advantage.

Dynamic entity is quite feasible using a differentiated approach to the workforce as a complex. This approach opens great prospects for ensuring that the organization's resources match changing strategic needs.

Research into industry prospects inevitably concerns the way the abilities, values and motivational attitudes of different groups of employees correspond and support the organization development.

The HR ecosystem must reflect not only this complexity, but also manage it, supporting diversity and integrity.

References

1. Becker, B.E., Huselid, M.A.: Strategic human resources management: where do we go from here? J. Manag. **32**(6), 898–925 (2006)
2. Mardini, G.H., Lahyani, F.E.: Impact of firm performance and corporate governance mechanisms on intellectual capital disclosures in CEO statements. J. Intellect. Cap. **23**(2), 290–312 (2022). https://doi.org/10.1108/JIC-02-2020-0053
3. Fulmer, I.S., Ployhart, R.E.: Our most important asset: a multidisciplinary/multilevel review of human capital valuation for research and practice. J. Manag. **40**(1), 161–192 (2014). https://doi.org/10.1177/0149206313511271
4. Schuler, R.S., Jackson, S.E.: Linking competitive strategies with human resource practices. Acad. Manag. Exec. **1**, 207–219 (1987)
5. Snell, S., Swart, J., Morris, S., Boon, C.: The HR ecosystem: emerging trends and a future research agenda. Hum. Res. Manag. **62**(1), 5–14 (2022). https://doi.org/10.1002/hrm.22158
6. Donnelly, R., Hughes, E.: The HR ecosystem framework: examining strategic HRM tensions in knowledge-intensive organizations with boundary-crossing professionals. Hum. Res. Manag. **62**(1), 79–95 (2022). https://doi.org/10.1002/hrm.22115
7. Zaharov, D., Lobacheva, A.: Organization HR-ecosystem formation and development. Manag. Pers. Intellect. Res. Russ. **9**, 23–26 (2020). https://doi.org/10.12737/2305-7807-2020-23-26
8. Yalenios, J., d'Armagnac, S.: Work transformation and the HR ecosystem dynamics: a longitudinal case study of HRM disruption in the era of the 4th industrial revolution. Hum. Res. Manag. **62**(1), 55–77 (2022). https://doi.org/10.1002/hrm.22114
9. Ginting, H., Khristiningrum, V., Gustomo, A., Wisesa, A., Saputra, J.: Dynamics of human resource department ecosystem in developing human resource role: an ecosystem perspective. PLOS ONE **18**(12), e0295544 (2023). https://doi.org/10.1371/journal.pone.0295544
10. Dharmasiri, A.: Exploring an HR Ecosystem: a Glimpse through Ten Gs. Kelaniya J. Hum. Res. Manag. (2015). https://doi.org/10.4038/kjhrm.v10i1-2.18

11. Ogiy, O.G., Osipov, V.Y.: Neural Network Forecasting of Labor Potential of Fishing Vessel Crews, Mar. Intel. Technol. **4**, 214–223. (2023). https://doi.org/10.37220/MIT.2023.62.4.026
12. Malik, A., Budhwar, P., Mohan, H., Srikanth N.R.: Employee experience –the missing link for engaging employees: Insights from an MNE 's AI -based HR ecosystem. Hum. Res. Manag. **62**(1), 97–115 (2022). https://doi.org/10.1002/hrm.22133
13. Singh, A., Pandey, J.: Artificial intelligence adoption in extended HR ecosystems: enablers and barriers. an abductive case research. Frontiers Psychol. **14**, 1339782 (2024). https://doi.org/10.3389/fpsyg.2023.1339782
14. Mesoeconomics of Development. Edit. by G.B. Kleiner. Moscow, Nauka Publ., 2011, 802 p.
15. Kleiner, G.B., Rybachuk M.A., Karpinskaya V.A.: Strategic Planning and Systemic Optimization of the National Economy. Stud. Russ. Econ. Dev. **33**(3), 243–248 (2022). https://doi.org/10.1134/S1075700722030054
16. Polterovich, V.M.: On the guide for reformers: some conclusions from the theory of economic reforms **1**(28), 7–24 (2005). (In Russ.)
17. Witt, U.: The role of human action motivation in explaining and evaluating economic evolution. Heterodoxia Versus Economic Reductionism: Micro-, Meso-, Macro-: ed. by S.G. Kirdina-Chandler, V.I. Maevsky. Moscow: IE RAS. 56–69 (2018). (In Russ.)
18. Prigozhin, A.I.: Goals and Values: New Methods of Work with the Future. Eds. 2nd. Moscow, Lenand 440 (2021) (In Russ.)
19. Kleiner, G.B.: System Economy: Development Steps: Monograph. – Scientific Library Publishing House. Finace Bus. 746 (2021)
20. Kleiner, G.B., Rybachuk, M.A.: System structure of the economy: qualitative time-space analysis. Fronteira J. Soc. Technol. Environ.Sci. **5**(2), 61–81 (2016)
21. Kooij, D., Van den Broeck, A.: Historical and Contemporary Perspectives on Work Motivation in Human Resource Management. In book: Oxford Research Encyclopedia of Business and Management Publisher: Oxford University Press (2022). https://doi.org/10.1093/acrefore/9780190224851.013.164
22. Kokurin, A.V., Ekimova, V.I., Doronin, V.Yu., Efanov, A.S.: Relationship of value-semantic attitudes and motivation of volunteers-rescuers. Hum. capital **9**(141), 283–287 (2020). https://doi.org/10.25629/HC.2020.09.27
23. Tishchenko, N.A., Luchinkina, A.I.: Features of the study of the socio-psychological component of organizational culture. Hum. capital **8**(152), 52–60 (2021). https://doi.org/10.25629/HC.2021.08.06
24. Salvador-Gómez, A., Llusar, J. Beltrán-Martín, I. A multi-actor perspective on the effectiveness of human resource management implementation: an empirical analysis based on the ability-motivation-opportunity framework. Int. J. Hum. Res. Manag. **34**(20), 3963–4002 (2023). https://doi.org/10.1080/09585192.2022.2163855
25. Yusup, A., Maulani, I.: The influence of transformational leadership on employee motivation. Devotion J. Res. Commun. Serv. **4**(3), 798–802 (2023). https://doi.org/10.36418/devotion.v4i3.441
26. Ogiy, O.G., Egorov, I.O., Kolyada, A.V.: System of Management of Labor Potential of the Fishery Complex. Software Registration Certificate No. 2023617705. Date of registration in the register FSIP RF: 12.04.2023 (In Russ.)
27. Ogiy, O., Osipov, V.: A compositional approach to labor potential evaluation and a neural network model for its forecasting. In: Polyakov, R. (eds.) Ecosystems Without Borders 2023, EcoSystConfKlgtu 2023. LNCS, vol. 705. Springer, Cham (2023). https://doi.org/10.1007/978-3-031-34329-2_15
28. McCrae Robert, R., Costa, Jr., Paul, T.: Understanding persons: from stern's personalistics to five-factor. Pers. Individ. Differ. **169**, 109816 (2021). https://doi.org/10.1016/j.paid.2020.109816

29. Schwartz, S.: Universals in the content and structure of values: theory and empirical tests in 20 countries. Adv. Exp. Soc. Psychol. **25**, 1–65 (1992). https://doi.org/10.1016/S0065-2601(08)60281-6
30. Gerchikov, V.I.: Typological concept of labor motivation. Motiv. pay **2**, 53–62 (2005). (In Russ.)
31. Weed, D.L.: Weight of evidence: a review of concept and methods. Risk Anal. **25**(6), 1545–1557 (2005)
32. Huselid, M.A., Becker, B.E.: Bridging micro and macro domains: workforce differentiation and strategic human resource management. J. Manag. **37**(2), 421–428 (2011) https://doi.org/10.1177/0149206310373400
33. Grjibovski, A.M.: Analysis of nominal data (independent observations). Hum. Ecol. **6**, 58–68 (2008)

Analysis of the Effectiveness of Ecosystems in Russia Based on Mathematical Methods Modeling and Forecasting

Artem Oleynikov[1](✉), Dmitriy Knyazev[1], Ekaterina Kharlamova[1], Sergey Sazonov[1], and Andrey Mamaev[2]

[1] Volgograd State Technical University, Volgograd, Russia
artemoleynikov36@gmail.com
[2] LLC LABTON, Volgograd, Russia

Abstract. The development of ecosystems, today, is of great importance for the socio-economic development of regions and countries. Given the rapid pace of change and uncertainty inherent in the modern economy, the development of business ecosystems can help entrepreneurs and managers predict and adapt to changes in the external environment, which is a key factor for success in modern business. The article explores the theoretical and methodological foundations of ecosystem creation and their similarity to biological species. Currently, there is a lack of research related to the analysis of the effectiveness of ecosystems. In this paper, the classification of ecosystems is considered, an analysis of the effectiveness of the leading Russian ecosystems based on the thermal correlation map is proposed. Based on the analysis, key indicators used to assess the performance of ecosystems were identified, such as the number of customers, profitability, number of services, etc. Russian companies were identified as leaders among ecosystems in terms of the "number of services" indicator.

Keywords: Ecosystem · Digital Platform Ecosystems · Correlation Heat Map · Random Forest · Decision Tree · Platform Ecosystems · Business Ecosystems

1 Introduction

Currently, in Russia, the growth in the pace of development and introduction of innovative goods and services to the market, the rapid development of digital technologies is putting forward new and increased requirements for models and ways of doing business. In recent years, there has been a significant increase in interest in the latest, modern business models based on deep collaboration of companies with each other to obtain synergetic effects from collaboration. It is well known that trading enterprises play a huge role in the Russian economy, providing the country's population with goods and services. Thanks to it, a significant share of revenues to regional and federal budgets is generated, namely, a tenth of all tax revenues to the budget of our country.

Ecosystem is a new business format that is based on the cooperation of companies in the exchange of information and the creation of new opportunities. It allows companies

R. Polyakov (Ed.): EcoSyst 2023, LNNS 1092, pp. 194–207, 2024.
https://doi.org/10.1007/978-3-031-67354-2_21

to enter the market with minimal costs and offers an effective combination of resources and innovative proposals. Participation in the ecosystem brings several advantages, as it unites companies of different profiles, satisfies a wide range of needs, and creates a comfortable environment for consumers, which increases the income of companies. Ecosystems integrate mechanisms that ensure a simple transition from service to service, increasing the level of user comfort. Companies creating such an innovative environment no longer focus only on the product, but strive to build a single customer route, which significantly increases their chances of attracting and retaining the maximum number of consumers.

2 Historical Aspects in the Development of Ecosystems

The term business ecosystem originated historically in the 1960s. Economic biology has been developed in such new concepts for economic science as the life cycle of goods, organizations, organizational pathologies, etc. Since that time, the concepts of "business ecosystem", "entrepreneurial ecosystem" have become widely used in both scientific and business communities. The time of the emergence of the theory of business ecosystems coincided with the active development of the Internet by business, the digitalization of the world economy, the acceleration of information exchange processes and the renewal of innovations [1–3].

In his work, D. Moore noted that the company should not be considered as an element of the industry, but as part of a business ecosystem that belongs to several industries (Table 1). Within it, companies jointly create opportunities based on innovation: they work cooperatively, while competing, to maintain new products and meet customer needs and, ultimately, launch a new round of innovation. The elements of the ecosystem evolve in such a way as to best meet the needs of the market. Since the publication of the article D. Mura's term "ecosystem" has expanded the scope of its application in business [4].

In total, according to the theory of J. Mura, for the survival of a business organization, it is necessary to manage the entire ecosystem of its habitat, including communications, as it interacts with other biological entities and the business environment.

Thus, the ecosystem of any organization consists of the organization itself, its suppliers and consumers, competitors, and stakeholders. It also includes state, municipal, and public organizations that interact with each other through competition and cooperation. The term "ecosystem" was transferred from biology to economics, retaining its structural components. A biological ecosystem consists of censes (communities of organisms), biotopes (habitats) and communications that ensure the flow of energy and substances. Researchers continued to develop the theory of business ecosystems, realizing the similarities between business organizations and biological species in terms of survival, adaptation, evolution, and the limitations of the life cycle.

They concluded that business systems are a kind of biological species. One of the main problems that continued to occupy the minds of researchers was the life expectancy of a business and its dependence on external factors, as well as the development of rules of adaptability for a successful existence.

Table 1. Concepts of ecosystems from the point of view of various authors

Author	Term	Resource
A. Tensley	From the point of view of biological science, an "ecosystem" is a set of compactly existing living organisms in relation to their environment, an ecosystem is in constant energy exchange with the natural environment, that is, it is an open system	The works of A. Tensley (1930s)
D. Moore	A business ecosystem is an economic community that is supported by the foundation of interacting organizations and individuals and is the body of the business world. In accordance with the definition of J. According to Mura, business ecosystems are dynamic and jointly developing communities consisting of diverse entities that create and receive new content in the process of both interaction and competition	"Predators and prey: a new ecology of competition in May/June 1993"
Deloitte Touche	Ecosystems are dynamic and jointly developing communities consisting of diverse entities that create and receive new content in the process of both interaction and competition	"Business ecosystems are coming of age"

Source: based on [2, 4, 5]

3 Analysis of the Activities of Russian and World Ecosystems

Many researchers in the field of digital economy note the need to form business ecosystems in connection with the development of digital technologies and platform-based business. It is important not only to have access to a variety of data, but also to be able to extract value from them and use it in the interests of business and society. However, it is becoming increasingly difficult to do this alone, so the economy of the future is focused on cooperation between various participants, changing value chains and the formation of new ecosystems.

Digital transformation contributes to the development of a business ecosystem where there are no restrictions on physical location, which opens unlimited opportunities for interaction with various suppliers and customers. Today, managers are changing their focus: instead of focusing only on their business, they are looking for new approaches to cooperation with transport companies, payment systems and other industry participants, as well as companies from related industries. Many companies see the potential in creating business ecosystems [6].

The range of types of economic ecosystems is expanding and today such ecosystems are included in the species composition (shown in Fig. 1) as [7]:

- innovative - the association of companies, organizations, educational institutions, and government institutions to create new ideas, technologies and products based on innovation clusters that unite companies in the same industry or area.
- venture - groups of investors, startups, accelerators, and incubators that provide financing and expert support for young companies.
- digital platform ecosystems - a set of companies, developers, investors, and users who interact with each other based on digital platforms that provide access to new technologies, products, and services, as well as contribute to the development of the digital economy.
- household ecosystems - a set of individual consumers who interact with each other since digital platforms related to services, purchases, information exchange and experience.
- national ecosystems - all ecosystems of households in a particular country or region that interact with each other and with other ecosystems, such as ecosystems of business, the state, and the public.

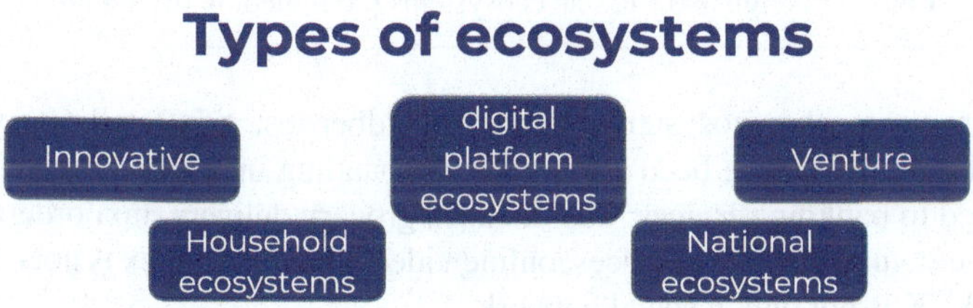

Fig. 1. Types of ecosystems [compiled by the authors], based on materials [7]

Figure 2 shows the different ecosystem levels by scale of activity offered by Gartner.

Fig. 2. Ecosystem levels [compiled by the authors], based on materials [7]

Modern ecosystems are currently actively developing in Russia and abroad. Global business ecosystems are complex and interconnected networks of companies, organizations and consumers that cooperate and interact to achieve their goals and ensure the sustainable functioning of the global economy. The first business ecosystems began to appear in the early 2000s, one of them was the digital ecosystem of Amazon. In the Russian market, Sber became a pioneer in creating a new business format, which in 2017 began to transform into an ecosystem. We can note that now, the leading ecosystems in Russia are: VTB, X5 Group, Sber, Tinkoff, Yandex, VK and Wildberries [3, 8]. Figure 3 shows advanced ecosystems that have been steadily entrenched in the Russian and global economy.

Fig. 3. Foreign and Russian ecosystems [compiled by the authors]

In recent years, the ecosystems of Sber, Wildberries, X5 Retail Group, Yandex, Tinkoff, VK and MTS have been significantly expanding in Russia. Sber, for example, is not limited to banking services: this includes grocery delivery, taxi ordering service, carsharing, e-commerce marketplaces, online video hosting. Yandex is not only a search engine, and VK is not only a social network.

Different ecosystems can vary greatly in size, composition, and principles of education. Often, in the context of ecosystems, they talk about digital giants, such as Sber, Tinkoff and Yandex, which form ecosystems at the industry level. The success of ecosystem development depends on the ability to gather the necessary knowledge, create knowledge assets from them and build a system of multi-level services around this core of key value. A business that creates and manages a key value attracts partners, offers the consumer a variety of related directions and thereby retains it. Interest for the studies provide these ecosystem solutions transaction type implemented by the IT giants as "Yandex" and Google, which offer their clients a variety of services and ecosystem, the industry grew out of brick-and-mortar companies, such as "Miratorg", "Ascona", X5 Groups (Fig. 4) [9].

In total, we can conclude that now the companies Sber and Yandex have the greatest strength against the background of the rest of the ecosystem sector in Russia. However, to consider, the peculiarities of Russian ecosystems, it is proposed to consider the ecosystem using the thermal correlation method. This method allows you to create a clearer understanding of the effectiveness of a particular ecosystem, including various approaches to the formation of ecosystems.

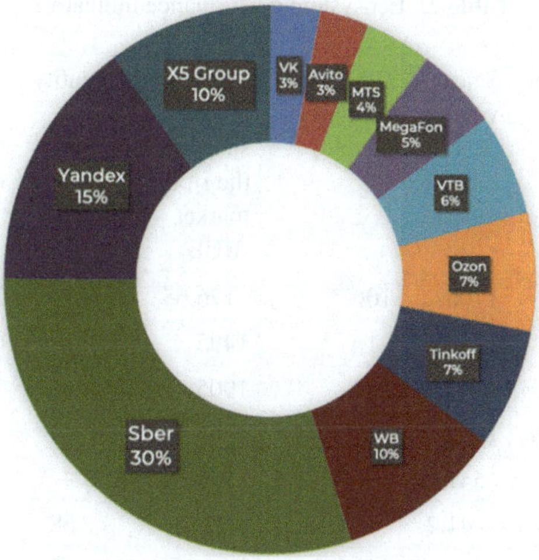

Fig. 4. Distribution of ecosystem strength in Russia (in %), based on materials: [8–10]

4 Application of the Correlation Heat Map Method to Analyze the Effectiveness of Russian Ecosystems

Let's consider the concept of efficiency for further analysis of the Russian segment of ecosystems: Efficiency is the effectiveness of the functioning of the system and the management process as the interaction of the managed and managing systems, that is, the integrated result of the interaction of management components. In a more simplified version, this definition can be presented as follows: efficiency is a comparison of the resources consumed with the result obtained. In assessing the effectiveness of management, the most difficult is to determine its result. To determine the effectiveness of ecosystems, the following indicators were identified (Table 2): net profit, sales volume, customers, the cost of one share on the stock market, quantity of services, the duration of the company's existence and profitability. In the analysis of the activities of the leading Russian business ecosystems, the correlation heat map method can be applied.

After conducting a primary analysis of the data, we built a heat map of correlations of features. The correlation was calculated using the Pearson method (Fig. 5). The map shows that the increase in the number of services in the ecosystem adversely affects the net profit of the company:

The Pearson correlation coefficient is based on the calculation of the covariance between two variables:

$$r = \frac{\sum (x_i - \overline{x})(y_i - \overline{y})}{\sqrt{\sum (x_i - \overline{x})^2 \sum (y_i - \overline{y})^2}}, \tag{1}$$

where x_i, y_i – i is the indicator, $\overline{y}$, $\overline{x}$ – is the average value.

Based on the data, a nonlinear model of forecasting the profitability of the company based on the "Decision Tree" algorithm was built. A decision tree is a method

Table 2. Ecosystem performance indicators

Company name	Net profit (BN rubles)	Sales volume (BN rubles)	Customers (million people)	The cost of one share on the stock market (RUB)	Quantity of services	Years on the market	Profi-tability (%)
Sber	270.5	1700	106.7	176.65	180 +	2	5.2
X5 Group	47513	1250	16	1495	4	6	7.3
Yandex	47.6	521.7	19	1905	90 +	5	12.3
Magnit	51.69	1856	63	4546	5	2	7
MTS	32.6	541	13.5	256.25	50 +	3	6
VK	41197	41.2	79.6	505.8	68	1	18
Wildberries	14.06	1680	120	-	33	3	23.3
Tinkoff	12	46.8	26.3	3992	50 +	1	7.7

Source: based on [11–16]

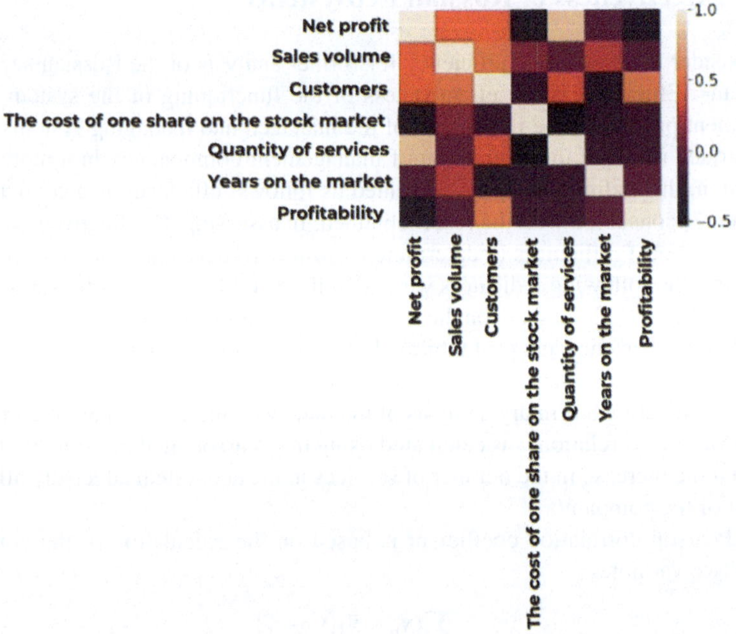

Fig. 5. Heat map of correlation according to the Pearson method [compiled by the authors]

of representing decision rules in a hierarchical structure consisting of two types of elements—nodes and leaves. The decision rules are in the nodes and the examples are checked for compliance with this rule by some attribute of the training set (Fig. 6). In the simplest case, because of the check, the set of examples that fall into the node is divided into two subsets, one of which contains examples that satisfy the rule, and the other does not. The standard error was chosen as the criterion for splitting the data [17].

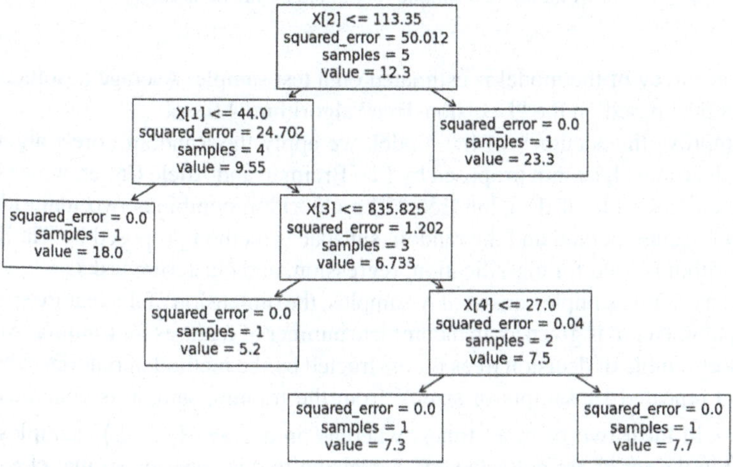

Fig. 6. Diagram of the "Decision Tree" algorithm for this model [compiled by the authors]

To build a forecasting model, the column "Profitability of data" was selected by the target function. All data except the column assigned as the objective function were taken as features. It is also necessary to divide the data into training (Fig. 7) and training samples (Fig. 8). The data were divided considering the uniform distribution of the target function in the training sample, i.e., it is necessary that both low-profit and high-profit companies were present in this sample [18].

	Company name	Net profit	Sales volume	Customers	The cost of one share on the stock market	Quantity of services	Years on the market
0	Sber	270.500	1700.0	106.7	176.55	180	2
1	X5 Group	47.513	[illegible]	16.9	[illegible]	4	8
5	VK	41.197	41.2	79.6	505.80	60	1
6	Wildberries	14.060	1680.0	120.0	NaN	33	3
7	Tinkoff	12.000	46.8	26.3	3992.00	50	0

Fig. 7. Training sample [compiled by the authors]

To assess the accuracy of the constructed model, a metric was adopted – the average absolute error:

$$\Delta = \frac{\sum(y_i - \hat{y})}{n}, \quad (2)$$

where y_i – the true value of the predicted value, $\hat{y}$ – predicted value, n – sample size.

	Company name	Net profit	Sales volume	Customers	The cost of one share on the stock market	Quantity of services	Years on the market
2	Yandex	47.60	521.7	19.0	1905.00	90	5
3	Magnit	51.69	1856.0	63.0	4546.00	5	2
4	MTS	32.60	541.0	13.5	256.25	50	3

Fig. 8. Test sample [compiled by the authors]

The accuracy of the model is estimated on a test sample. Average absolute error Δ for this model based on the "Decision Tree" algorithm: 4.089%

To improve the accuracy of the model, we apply the Random Forest algorithm, a machine learning algorithm proposed by Leo Breiman and Adele Cutler, which consists in using an ensemble of decision trees. The algorithm combines two main ideas: the Breiman bagging method and the random subspaces method proposed by Tin Kam Ho. The algorithm is used for classification, regression, and clustering tasks.

Let the training sample consist of *N* samples, the dimension of the feature space is *M*, and the parameter m is given as an incomplete number of features for training. At the first stage, an ensemble of decision trees is constructed by the method of random subspaces – a random repeated subsample of size *N* from the training sample is generated. Some samples will hit it two or more times, whereas on average $\left(1-\frac{1}{\mathrm{N}}\right)^2$ samples are not included in the set or are not selected. A decision tree is constructed that classifies the samples of this subsample, and during the creation of the next node of the tree, the algorithm selects a set of features based on which the partition is performed (not from all *M* features, but only from m randomly selected ones). The selection of the best of these *m* features is carried out by minimizing the mean square deviation:

$$\sigma = \sqrt{\frac{\sum(x_i-\overline{x})^2}{n-1}} \tag{3}$$

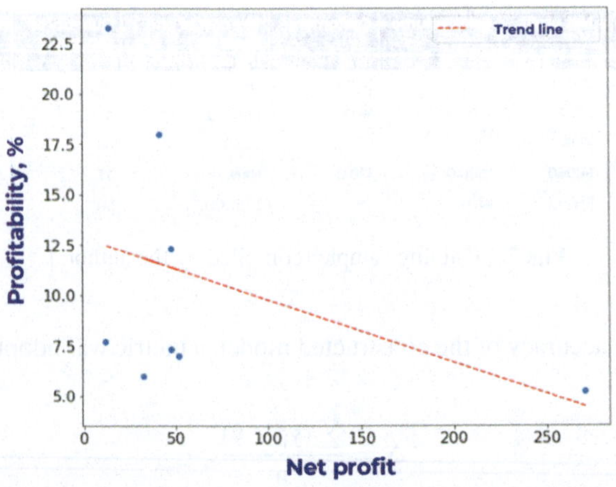

Fig. 9. Dependence of profitability on net profit [compiled by the authors]

Each tree makes a forecast, and the final value of the predicted value is the average value between the forecasts of the trees. The optimal number of trees is selected in such a way as to minimize the error of the algorithm on the test sample. 100 decision trees were adopted for this model. The average absolute error of the model forecast was 3.457%. To improve the result, it is also necessary to investigate the features of this model. Having plotted the dependence of profitability on net profit, with a net profit of less than 100, the distribution density is the highest (Fig. 9) [17].

Let's introduce a new feature "Net profit less than 100". Having also plotted the dependence of profitability on the number of customers, a high distribution density is observed with the number of customers less than 40 (Fig. 10). Let's introduce this feature.

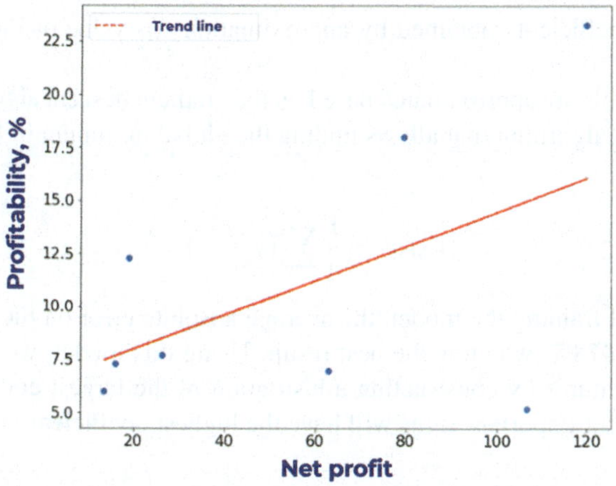

Fig. 10. The dependence of profitability on the number of customers [compiled by the authors]

The data for modeling will have the form (Fig. 11):

	Company name	Net profit	Sales volume	Customers	The cost of one share on the stock market	Quantity of services	Years on the market	Profitability, %	Customers is less than 40	Net profit is less than 100
0	Sber	270.500	1700.0	106.7	176.55	180	2	5.2	0	0
1	X5 Group	47.513	1250.0	16.0	1495.00	4	6	7.3	1	1
2	Yandex	47.60	521.7	19.0	1905.00	90	5	12.3	1	1
3	Magnit	51.69	1856.0	63.0	4546.00	5	2	7.0	0	1
4	MTS	32.60	541.0	13.5	256.25	50	3	6.0	1	1
5	VK	41.197	41.2	79.6	505.80	68	1	18.0	0	1
6	Wildberries	14.060	1680.0	120.0	NaN	33	3	23.3	0	1
7	Tinkoff	12.000	46.8	26.3	3992.00	50	0	7.7	1	1

Fig. 11. Modified data table [compiled by the authors]

A linear regression algorithm is used to build the model. Data splitting is like the examples above. It is also necessary to normalize these features for an adequate assessment of regression coefficients. Normalization occurs in the following way:

$$t = \frac{x_i - \mu}{\sigma}, \tag{4}$$

where x_i – the current sample, μ – the mean value for the trait, σ – the standard deviation for the trait. Thus, the values of all the signs will be in the range from 0 to 1.

The analytical record of the constructed regression has the form:

$$\hat{y} = a_0 + a_1x_1 + a_2x_2 + \cdots + a_nx_n, \tag{5}$$

where a_i– the coefficients obtained by approximation, x_i- value of the corresponding attribute.

The coefficients are approximated based on the gradient descent algorithm, a numerical optimization algorithm that allows finding the global minimum of the loss function L_{MSE} :

$$L_{MSE} = \frac{1}{n}\sum(y_i - \hat{y})^2 \tag{6}$$

As a result of training the model, the average absolute error on the test sample was obtained $\Delta = 2.478\%$, which is the best result. Using this model, we will analyze the importance of features by constructing a histogram of the largest coefficients modulo (Fig. 12). The most important signs will have the highest coefficient [19, 20].

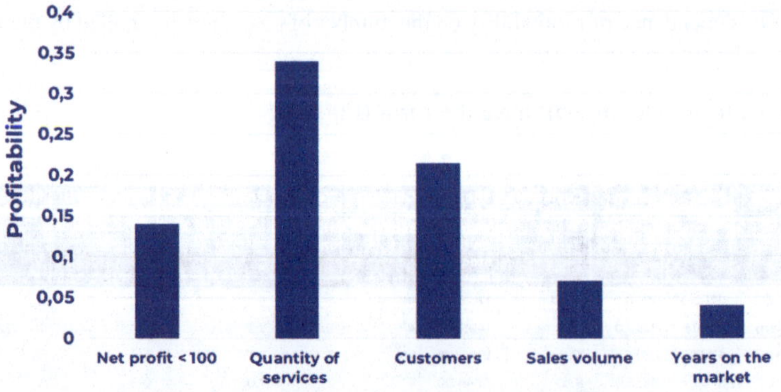

Fig. 12. Histogram of regression coefficients [compiled by the authors]

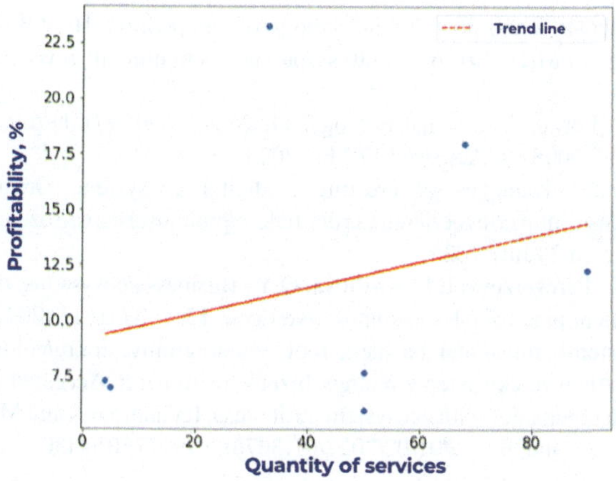

Fig. 13. Dependence of profitability on the number of services in the ecosystem [compiled by the authors]

5 Conclusions

Analyzing the results of the model based on linear regression, we can conclude that complex nonlinear algorithms do not always give the best result for predicting a continuous value on a small sample, linear algorithms do better. It can also be concluded that the number of services in the ecosystem has the greatest impact on the profitability of the company: the more services, the more profitable the company is (Fig. 13). It should be noted that in a comprehensive and systematic analysis of the effectiveness of ecosystems, it is important to consider all ecosystem indicators in the aggregate to obtain the correct result. The result of our research is that the effectiveness of ecosystems can be analyzed by the following indicators: net profit, sales volume, number of customers, the value of shares on the stock market, the number of years of existence on the market and profitability, but the key is the number of services in the ecosystem. Considering all the identified indicators and applying the considered mathematical methods, it is possible to analyze the effectiveness of any ecosystem in the future. To date, according to the indicator "number of services", the ecosystem of the company "Sber" is the leader of the ecosystem market, and therefore modern ecosystems need to open new services to develop successfully in the modern economy to develop their activities.

References

1. Ecosystem architecture. History of business ecosystem development. habr.com/ru/companies/nspk/articles/532462/. Accessed 9 May 2023
2. Business ecosystems as a modern market trend / Colvir Company, mis.business/upload/iblock/7ea/7ea49ac60a191d47ea71dd89a2d1eced.pdf. Accessed 13 May 2023
3. The history of the development of the first ecosystem in Russia. vc.ru/u/969260-innovation-agent/333608-istoriya-razvitiya-pervoy-ekosistemy-v-rossii. Accessed 9 May 2023

4. Moore J.F.: Predators and prey: a new ecology of competition. Harv Bus Rev. 1993 May-Jun;71(3):75–86. PMID: 10126156, https://pubmed.ncbi.nlm.nih.gov/10126156/, Accessed 28 April 2023
5. WILLIS, A.J.: Forum. Functional Ecology, **11**, 268–271 (1997). https://doi.org/10.1111/j.1365-2435.1997.00081.x. Accessed 17 May 2023.
6. Volodina N.L.: Advantages of creating a digital ecosystem. Organizer of production. (2021). №4., https://cyberleninka.ru/article/n/preimuschestva-sozdaniya-tsifrovoy-ekosistemy. Accessed 28 July 2023
7. Kulapov, M.N., Pereverzeva, E.I., Kirillova, O.Y.: Business ecosystems: definitions, typologies, development practices. Issues Innovative Econ. **12**, 1597–1612 (2022)
8. Digital ecosystems are a real business tool. www.agroinvestor.ru/column/maksim-nikitochkin/32980-tsifrovye-ekosistemy-realnyy-biznes-instrument/. Accessed 28 March 2023
9. Experts named companies with ecosystem attributes / Technologies and Media, https://www.rbc.ru/technology_and_media/01/02/2022/61f3d76f9a794775ff544309. accessed 9 April 2023
10. Business ecosystems: concept, advantages, forecasts, sales-generator.ru/blog/biznes-ekosistemy/. Accessed 07 May 2023
11. Opportunities of the X5 Group ecosystem, dialog.x5.ru. Accessed 28 May 2023
12. Official website of PJSC Magnit, https://www.magnit.com/ru/about-company/about-magnit/. Accessed 26 May 2023
13. Forecast of the profitability indicator of the company "Sber" by the end of (2022). http://www.finmarket.ru/news/5909917. Accessed 15 April 2023
14. VK's financial and operational results for the third quarter of (2022). https://vk.company/ru/investors/info/11320/. Accessed 14 March 2023
15. The price of ecosystem shares in Russia (OJSC Magnit (MGNT)), https://ru.investing.com/equities/magnit_rts. Accessed 14 March 2023
16. MTS net profit by the end of (2022). www.kommersant.ru/doc/5861186. Accessed 28 March 2023
17. Mitina, T. V. Multidimensional random variables. Correlation analysis: a textbook / T. V. Mitina. — Dubna: Dubna State University (2022). https://e.lanbook.com/book/196978. Accessed 15 May 2023
18. sklearn.tree.DecisionTreeRegressor — scikit-learn 1.2.2 documentation, scikit-learn.org/stable/modules/generated/sklearn.tree.DecisionTreeRegressor.html. Accessed 07 May 2023
19. sklearn.ensemble.RandomForestRegressor — scikit-learn 1.2.2 documentation, scikit-learn.org/stable/modules/generated/sklearn.ensemble.RandomForestRegressor.html. Accessed Accessed 10 May 2023
20. Rashka, Sebastian, Mirjalili, Vahid. Python and machine learning: Machine and deep learning using Python, scikit-learn and TensorFlow 2, 3rd ed.: Translated from English – St. Petersburg: Dialectics LLC (2020)
21. Implementation of the ecosystem approach as a vector of sustainable socio-economic development of the territory: monograph / S.P. Sazonov, I.A. Ezangina, E.E. Kharlamova, I.A. Chekhov, A.A. Hryseva. Kursk (2022)
22. Petrova, Y.S.: Modern trends in the development of digital ecosystems in the financial sphere / Petrova, Y.S., Kharlamova, E.E., Rzaeva, I.F. Manag. Acc. **12** (3), 765–771 (2022)

23. Problems and prospects of development of business ecosystems in Russia / A.A. Oleynikov, E.E. Kharlamova. Management and finance of production systems : collection of scientific and practical. Scientific and practical conference (Volgograd, December 12, 2022) / ed.: I. A. Ezangina, I. A. Chekhov ; editorial board: S. P. Sazonov [et al.] ; Volgograd State Technical University, Volgograd Institute of Management – branch of RANEPA under the President of the Russian Federation, Volgograd branch of Plekhanova: 149–153 Kursk (2023)
24. A Look Inside the New Trends in Business / THE FUTURE OF WORK – Fast Company, https://www.fastcompany.com/3045104/a-look-inside-the-new-trends-in-business. Accessed 15 April 2023

The Conceptual Content of Cybersecurity Ecosystems: A Professional Language Perspective

Irina Guseva and Elena Pliva(✉)

Kaliningrad State Technical University, Kaliningrad, Russia
Elena.pliva@klgtu.ru

Abstract. The article is devoted to the consideration of the features of metaphorical concepts in the ecosystem of "cybercrime" based on media texts. The concept of cybercrime is defined as a complex, integrated entity covering a wide range of criminal activities inherent in crimes in the real world, reflecting the global nature of this phenomenon and the necessary means to confront and combat it. The cognitive frame model of the concept "cybercrime" is presented and analyzed, as well as methods and mechanisms of linguistic representation of its metaphorical components. The results of the study give a clear idea of how the metaphor shapes our cognitive perception of the virtual world of cybercrime, the perception of technical achievements, helps to interact with them, overcome the problem of the impalpability and invisibility of the Internet, correlates our physical experience with the virtual, familiar with an unusual view of the world, forming a "digital space".

Keywords: Metaphorization · Conceptualization · Discourse · Metaphorical Model · Ecosystem · "It-Detective" · Cybercrime · Cybercriminals

1 Introduction

Intensively developing over the last decade, the Internet has deeply penetrated our lives, completely transforming everyday habits, work, study, and our perception of the world. It has given rise to professions that are in demand in a new, actively functioning "Internet habitat", in which the problem of cybersecurity becomes especially relevant. According to experts in the field of information security, the number of attacks using malware has increased dramatically recently, and the growth of activity in the cybercrime ecosystem continues. IT detectives and cyber criminologists are stepping up to protect the "digital world" and fight cybercrime. As Internet technologies continue to advance, so do the cybersecurity threats that accompany them. There is a need for effective ways to protect the "digital side" of our lives from criminal intrusions into it.

R. Polyakov (Ed.): EcoSyst 2023, LNNS 1092, pp. 208–214, 2024.
https://doi.org/10.1007/978-3-031-67354-2_22

2 Materials and Methods

The objectives and methods of the research are to identify the features of metaphorical modeling of situations related to the "Cybercrime" conceptual sphere in Russian and English media texts. Using conceptual and frame analysis, the study aims to consider ways of conceptualization and categorization of knowledge verbalized in discursive formations. The paper presents the results of research in this area, focusing on the most significant concepts included in this conceptual sphere and creating a metaphorical image of its participants and participating elements functioning in our perception of the Cybercrime ecosystem. In order to fully represent the meaning of a word, its cognitive frame model is used in the work as a structured unit of knowledge with interrelated components reflecting opinions, assessments, and definitions that stand behind each language unit [1, 2]. According to the theory of conceptual metaphor, the metaphor mechanism is the relationship between two knowledge structures: the cognitive structure of the original donor zone and the cognitive structure of the recipient zone [3].

3 Results

The concept of Cybercrime is considered a complex entity that encompasses all the attributes of a crime and represents the global nature of this phenomenon. Cybercrime "has merged into a wider criminal ecosystem" [4]. The concept of the cybercrime ecosystem is still developing, much like any other ecosystem. Cybercrime plays a central role in this system, representing its global nature and complexity. Cybercrime is regarded as a "peek at the cybercriminal ecosystem" [5]. "Cybercrime is a dynamic landscape, with the types of attacks and the nature of the operators or gangs involved becoming ever more sophisticated" [6]. Cybercrime is a well-organized business with a hierarchical structure, managed by professionals. The role of cybercriminals is played by ransomware gangs consisting of team leaders, malware developers, data collectors, etc.

The discourse in this area is filled with metaphorical images, behind which the Internet "hides and manifests itself" [7]. In the era of the "ubiquity" of smartphones and the Internet, metaphor forms our cognitive perception of technical achievements, helps us to interact with them, and overcome the problem of the intangibility and invisibility of the Internet. Metaphors help to bridge the gap between our physical experiences and the virtual world, making the unfamiliar aspects of the digital space more relatable and understandable. In the discourse of cybercrime, hacker attacks, data leaks, internet fraud, breaches, and ransomware incidents occur. There are threats, blackmail, and victims of crimes, as well as detectives, criminal cases, and investigations taking place. The cybercrime ecosystem itself can be represented schematically in the discourse under consideration (Fig. 1):

Within the framework of this study, our interest is focused on those elements of this ecosystem that are centered around metaphorical images, since it is through metaphors that we can understand how we interpret certain concepts.

Let's examine some of the key components of this ecosystem. Who are the **cyber detectives/investigators** who are at the forefront in the fight against cybercrime? They are "responsible for investigating and analyzing cybercrime incidents, such as hacking,

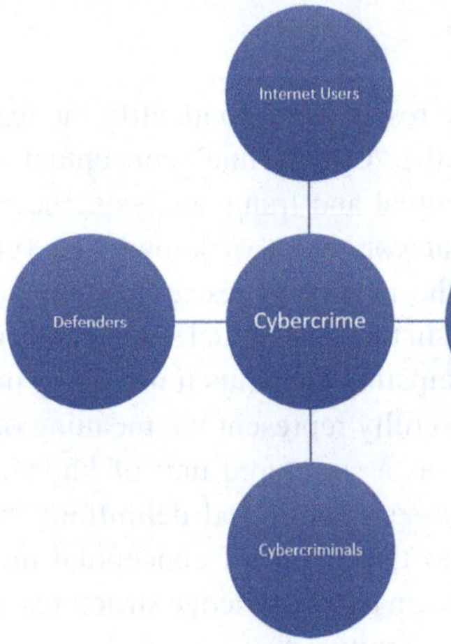

Fig. 1. Cybercrime ecosystem.

identity theft, fraud, and other types of cyber-related crimes" [8]. Another aspect that is related to this concept is the field of cybersecurity forensics: "Cybersecurity forensics is like being a digital detective. It involves the careful and systematic process of collecting, analyzing and preserving digital evidence to uncover what happened in a cybercrime or security breach [9]. This implies that the investigative process of a cyber detective consists of the same steps just like detective work in the physical world – data collection, data analysis, reporting, presentation, legal considerations. However, there are also significant differences in the object of crimes and methods inherent only in the Internet environment, mainly concerning computer software. An IT detective examines so called malware artifacts, or traces left by malware: "These artifacts include unusual files, changes in system settings, or the installation of new, unfamiliar software" [9].

As a detective in a real world, digital detective is armed with the following tools:

Footprinting. At its core the term footprinting conveys the idea that a company in the process of its operational activity inevitably leaves traces which can then be analyzed and used to assess the security posture. The metaphor behind this term is that just as footprints can give an idea of a person's movements and actions, a company's digital activities can give an idea of its security practices and vulnerabilities. In cybersecurity, footprinting means collecting sensitive information about your target system to determine how a successful system attack can occur [10]. This may include information about the target security system, situation, and mode of operation. Cybersecurity experts have been using footprinting for years to determine vulnerabilities in networks, systems, and IT infrastructures and develop appropriate protection measures.

Reverse Engineering. It is a process of taking a piece of software or hardware and analyzing the functions and information flow so that its functionality and behavior can be understood [11]. The metaphor behind the term "reverse engineering" in cybersecurity is that it involves taking apart a technology or system to understand how it works, much like reverse engineering a physical object to understand its design and function. Cybersecurity experts have been using reverse engineering techniques for a long time to analyze malware to dissect viruses and create protection tools. Reverse engineering is intrinsically tied to backhack, also known as "counter-hacking," a term used in cybersecurity to describe the act of retaliating against a hacker or cyber attacker by using their own tactics against them. This process involves reverse engineering, analyzing, and understanding the techniques used in cyberattacks. The image behind the term could be that of someone breaking into a system or network from the backdoor, gaining unauthorized access [12].

File carving is another valuable tool used by cybersecurity experts. It is a fascinating technique that allows to recover lost or damaged files. Он незаменим in digital forensics, where specialists have to check for any deleted or formatted files for further investigation, and to collect and restore evidence at the crime [13]. The metaphor behind it is that just as a sculptor carefully removes unwanted material to reveal the desired shape, file carving carefully extracts specific data from a larger file.

Patch. Initially it was a piece of material that is used to cover a hole in something or to provide extra protection to an area [14]. But in the context of programs, not only do they solve problems, but also make software better. Security patches address any discovered vulnerabilities that software companies identify. The patch consists of a set of changes for the code to fix bugs that could lead to cybersecurity incidents. The patch usually applies to one component of the software [15].

Honeypots. As it is known, the term honeypot, originally meaning something highly desirable, came from the sphere of espionage, where it was used in the meaning of a trap for stealing secrets. This method of finding out secrets involved the use of romantic relationships, the same honeypots, for compromising and then finding out secrets. In the field of cybersecurity, honeypots are a type of Internet security resources that are used as bait for attackers to encourage them to invade the network for any illegal use, while they do not carry any valuable data, being, in fact, a fake host [16]. The function of a honeypot is to represent itself on the internet as a potential target for attackers -- usually, a server or other high-value asset -- and to gather information and notify defenders of any attempts to access the honeypot by unauthorized users [17]: "There's a lot to be gleaned from watching a hacker interact with a honeypot - and you don't have to put the rest of your network in danger to do so" [18].

The aim of an IT detective/criminalist is to protect Internet users from malicious activities such as hacking, phishing, and spam in order to prevent them from becoming victims of cybercrime. Various types of viruses and worms spread across the network, stealing and infecting targeted programs and messages either independently or with the help of cybercriminals. Computer worms can spread from one computer to another on their own, while viruses require human assistance to spread. Malware acts in a similar way to a human criminal, embodying various aspects of the criminal world. This metaphorical transfer allows them to act as if they were in a physical environment, which enables us to define a metaphorical model of conceptualization of a malicious virus program as: A VIRUS IS A LIVING BEING. The metaphorization in language is based on "characteristics/function."

Virus can:

- "extort, blackmail": Ransomware; Cyber Extortion;
- "steal": Phishing;
- "capture": Botnetting;
- "intimidate": Cyber Bullying;
- "stalk/track": Cyber Stalking;
- "piraty": Software Piracy;
- "launder money": Electronic Money Laundering;
- "cheat": Scamming;
- "carry out terror": Cyber terrorism.

Let's consider such a hacker attack as **phishing**, in which personal data and money are stolen using fake sites. Phishing has its origins from the word "fishing," as scammers use bait to lure in their victims. They manipulate basic human emotions and use information that they could find in open sources about a person to influence them. A phishing email is often disguised as a request from a legitimate company, prompting the recipient to provide their personal data. The attacker's goal is to steal money, gain access to sensitive data and login information, or to install malware on the victim's device [19]. In the metaphor of "fishing," the bait is a phishing letter and the whole process resembles fishing with a fishing rod.

Trojan. The sophisticated and cunning engineering of the Greeks inspired the name for a group of malicious digital tools known as Trojan programs. These programs are designed to cause damage to the victim's computers unnoticed, by reading passwords, recording keystrokes, or downloading other malware that can take over the entire computer. Computer scientists aptly chose the term "Trojan horse" for these programs, as they enter the computer under the guise of harmless programs and begin to act after being launched. The Trojan's actions may go unnoticed by the victim, as it is often written into the source code of the program, making it indistinguishable from the original. This allows the virus to pursue the hacker's goals without detection [20]. The cunning of the Greeks has been interpreted into sophisticated viruses, forming the basis of the metaphorical name.

Cybercriminals can use **botnets** to send spam emails. A botnet is a network of hijacked Internet-connected devices on which malicious code is installed. Infected devices are known as bots, and a cybercriminal remotely controlling them is called a "**botnet herder**" by analogy with a shepherd herding a flock [21].

Ransomware is «a type of malicious software that threatens to publish the victim's data or perpetually block access to it unless a ransom is paid» [22]. In this case, a malicious program is likened to a human demanding a ransom.

4 Conclusion

The research findings have led to the development of a conceptual model of cybercrime, which includes cybercriminals, investigators, and victims as the main actors. The cognitive-discursive analysis of key concepts in this model has revealed the underlying knowledge behind the metaphorical terminology used in this area and made it possible to determine the source or the initial conceptual area of metaphorical comparison and to determine the parameters by which the transfer of meaning in the area under consideration takes place. In the realm of IT detective work, concepts such as "tool" and "method/means" are employed, transferring meaning from the physical to the digital realm. In the field of cybercrime, malicious activity is personified through the ability of virus programs to "move and spread, infect and attack" their victims. Two worlds, two camps in the conceptual field of cybercrime confront each other: IT defenders/detectives and ransomware gangs using their means of organized attack on targeted victims. Metaphorical terminology perfectly conveys the atmosphere of tension and aggression in the world of Cybersecurity and Cyber threats.

References

1. Arutyunova, N.D.: Metaphor and discourse. In: Arutyunova, N.D. (ed.) Theory of metaphor, pp. 5–33. Progress, Moscow (1990)
2. Boldyrev, N.N.: Cognitive semantics. Introduction to cognitive linguistics: a course of lectures. Tambov., TGU im. Derzhavina (2001)
3. Lakoff, G., Johnson, M.: Metaphors We Live by. Editorial URSS, Moscow (2004)
4. The digital detective: Mikko Hypponen's war on malware is escalating- Wired UK. https://www.wired.co.uk/article/the-digital-detective. (Accessed 21 Oct 2023)
5. Cybercrime - A Peek at the Cybercriminal Ecosystem | Institut Montaigne. https://www.institutmontaigne.org/en/blog/peek-cybercriminal-ecosystem. (Accessed 25 Oct 2023)
6. The Cybercriminal Ecosystem: Evolution and Extortion. https://www.lawsociety.org.uk. (Accessed 16 Feb 2023)
7. Schetvina, A. (2023) Serfers in a global net. How the Internet conceals and reveals itself through metaphors. https://supernova.is/editorial/metaphors. (Accessed 21 Oct 2023)
8. How to Become a Cybercrime Investigator (cybersecurityguide.org) https://cybersecurityguide.org/careers/cyber-crime-investigator. (Accessed 21 Nov 2023)
9. What is Cybersecurity Forensics? The Art of Being a Digital Detective. https://www.freecodecamp.org/news/what-is-cybersecurity-forensics/ (Accessed 24 Nov 2023)
10. What are footprinting and fingerprinting? Shorebreak Security - Experts in Information Security Testing. https://www.shorebreaksecurity.com/blog/what-are-footprinting-and-fingerprinting/(Accessed 20 Nov 2023)
11. Cyber Security Tools | Complete Guide on Top 5 Cybersecurity Tool (hkrtrainings.com). https://hkrtrainings.com/cyber-security-tools. (Accessed 24 Nov 2023)
12. What is Back-Hack? CellularNews. https://cellularnews.com/definitions/what-is-back-hack/(Accessed 21 Oct 2023)
13. What is File Carving? | CellularNews. https://cellularnews.com/definitions/what-is-file-carving/(Accessed 21 Oct 2023)
14. Patch Definition & Meaning. https://www.britannica.com/dictionary/patch (Accessed 21 Oct 2023)
15. Software vulnerability patching best practices: Patch everything, even if vendors downplay risks | Infosec (infosecinstitute.com). https://resources.infosecinstitute.com/topics/vulnerabilities/software-vulnerability-patching-best-practices-patch-everything-even-if-vendors-downplay-risks. (Accessed 21 Nov 2023)
16. Detailed Guide on Honeypot / Хабр (habr.com). https://habr.com/ru/companies/alexhost/articles/528796. (Accessed 21 Oct 2023)
17. What is a honeypot? How it protects against cyberattacks (techtarget.com). https://www.techtarget.com/searchsecurity/definition/honey-pot (Accessed 21 Nov 2023)

18. What is a honeypot? | Definition & explanation of how they work (proprivacy.com). https://proprivacy.com/guides/what-is-a-honeypot. (Accessed 21 Nov 2023)
19. What Is Phishing? Examples and Phishing Quiz - Cisco. https://www.cisco.com/c/en/us/products/security/email-security/what-is-phishing.html. (Accessed 24 Nov 2023)
20. Trojan. https://promopult.ru/library>Trojan. (Accessed 21 Nov 2023)
21. What Is a Botnet? (kaspersky.com). https://www.kaspersky.com/resource-center/threats/botnet-attacks. (Accessed 21 Nov 2023)
22. Ransomeware. https://glosbe.com/en/en/Ransomware(Accessed 21 Oct 2023)

Features of the Operation of Digital Ecosystems When Recording Security Events

Vladislav Podtopelny(✉) and Alina Babaeva

Kaliningrad State Technical University, Kaliningrad, Russia
ionpvv@mail.ru

Abstract. The aspects of creating a vector of security events of information systems are studied. The methods used in the construction of the vector of security events are considered, their features are characterized, the specifics of using Markov chains to simulate the attacking actions of the intruder. The suitability of these methods for determining the parameters of the vector is determined, taking into account the use of elements of artificial intelligence in identifying signs of security events of various types. When creating a vector security events take into account the specifics of determining the probabilities of system transitions to various states of compromise. To solve the problems of searching for security events and the formation of their vectors, it is proposed to study the problems of modular architecture. The features of creating a simplified vector of security events are considered, taking into account the links between the tactics (states) proposed in the FSTEC methodological documents.

Keywords: Ecosystem · Digital Eco-Environment · Resource Value · System Criticality · Subsystem · Threat · System Error

1 Introduction

When studying the information security of segmented automated systems, it is required to form vectors of security events for attackers. In accordance with the methods, the determination of security event vectors is based on the expert opinion of information security specialists [1]. Markov chains allow you to take into account the probabilities of system transitions to different states of segment compromise, which can be useful when creating a security event vector.

For a more accurate construction of the vector of security events and the definition of its parameters, it is necessary to take into account the specifics of the classification of the feature space of network anomalies and use formal methods for determining the sequence of tactics given in the methodological document. These methods should be simple and applicable in practice when building a threat model [2]. In addition, it is necessary to take into account the ambiguity of the signs of attacking influences, on the basis of which the vector is built. Accordingly, it is necessary to study the possibilities of mathematical and algorithmic support for searching, analyzing and predicting security events in segments of information systems with ambiguously interpreted signs of the occurrence of security events.

R. Polyakov (Ed.): EcoSyst 2023, LNNS 1092, pp. 215–220, 2024.
https://doi.org/10.1007/978-3-031-67354-2_23

2 Materials and Methods. Definition of Specifics

The security event vector must take into account the possibility of new events that are interpreted as dangerous. When constructing such a vector, it is necessary to take into account aspects that can change the direction of security events. This approach is especially useful when auditing multi-level digital systems, as it takes into account the peculiarities of the data transmission technologies used. The continuation of the vector or the definition of a new direction depends on the success of previous attack sequences.

There are the following methods of analysis applicable to the detection of security incidents: statistical analysis, wavelet analysis, cluster analysis, fractal analysis, support vector method, genetic algorithms, immune systems, neural networks, decision trees, Bayesian networks, expert systems. Statistical methods, due to incorrect selection of a set of observable parameters or incorrect sizes of generalization of source data sets, can lead to the fact that the model describing the flow of network traffic will register false alarms of the system or the inability to detect intrusions. The advantages of statistical methods are the relative clarity of the results, their adaptability to changes in the behavior of users of digital segments, as well as the ability to detect anomalous events [3].

Veivlet-analysis helps in the selection of coefficients of data on the signs of incidents, described by data sets, by basic functions. As a signal, you can consider the volume of network traffic per unit of time.

Cluster analysis of network traffic helps to identify such characteristics in digital segments that will allow you to find groups, among which signs of a normal and abnormal state will be highlighted. Neural networks are often used as a way to detect anomalies. However, they require a large amount of computing resources and training time. To determine the specifics of attacking influences, the use of Markov networks [4].

Events in digital ecosystems of typical processes are discrete. They use data from many services to work simultaneously. They act as aggregating or central elements in the architecture of the digital eco-environment. There are also processes that work individually with autonomous services of digital ecosystems, without implying the distribution of their data to other services.

3 Results. Determination of Importance

The digital ecosystem system involves several functional blocks for security management [5]. They must perform additional tasks to ensure that security events are logged and handled. The model includes four subsystems:

1. Threat data collection module.
2. Module for analyzing the log file of data on threat parameters and saving the analysis results to a text file.
3. An additional neural network-based analyzer or other analytical tool for network or other parameters, or additional pluggable components.
4. A consulting module integrated into the overall consultation subsystem.

All these components are combined into assemblies for the monitored subsystems (Fig. 1).

The following types of assemblies are presented:

1. In control and data collection systems.
2. In separate interfaces of control over automation objects.
3. In applications for the analysis of production and technological data.

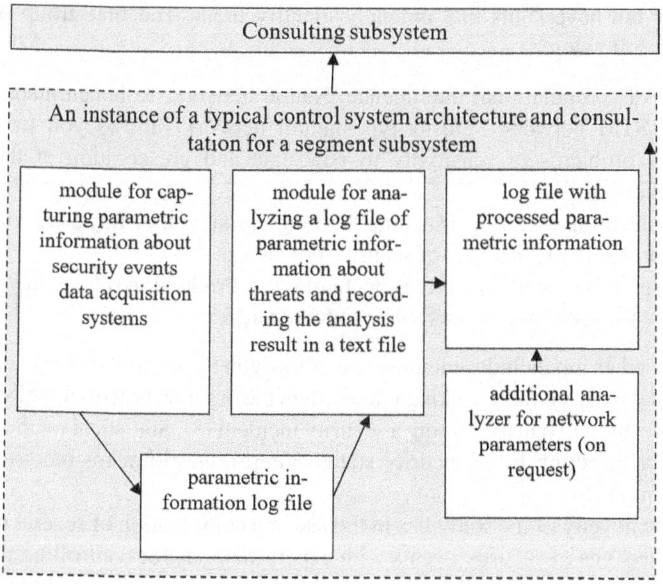

Fig. 1. The general architecture of the typical assembly of security event registration modules combined with digital segments

The module must collect parametric information, including from third-party systems. In addition, the module must work in conjunction with additional specialized modules that collect network modules. A sniffer, often used in such systems, writes information to a text file in a specific format that depends on the transport protocol. These analyzers can contain their own additional analyzers, the results of which will be taken into account in the production network of rules of the main analyzer. The Captured Traffic Log File Analysis and Parametric Data module analyzes anomalies and security events, and saves anomaly data to a text file. The module of consultation and selection of countermeasures performs the function of consulting to identify a specific threat or anticipated security events and gives recommendations to the user, and also provides the ability to run a connected external module to confirm the results of the analysis or for their use in other analytical processes [6].

The above architecture of the management and consulting system provides:

1. The ability to change the composition and functionality of the system.
2. Decentralized plug-in management.

The above architecture provides flexibility of functionality. It will allow you to detect and prevent possible security events much more efficiently than other systems. Thus, taking into account the specifics of the above studies, it is possible to cite the most promising methods and methods for determining attack vectors, which make it possible to solve the tasks set in the study of constructing security event vectors and predicting them [7]. Two groups of methods are used. The first group includes artificial intelligence methods that make it possible to determine the signs of the presence of a threat being implemented in the information system. Of particular note is the fact that the threats sought may not have signs that uniquely identify them. The first group of methods includes the following:

1. Methods of computational intelligence. Neural network. Kohonen maps, ART networks, LSTM networks, cluster-type neural network (allows you to solve contradictory problems of sensitivity to new data and preservation of the acquired knowledge).
2. Machine learning methods. Machine learning classifiers. A Bayesian trust network enables probabilistic analysis of security processes.
3. Knowledge-based methods. Expert technologies. Production rules. They are used to determine the sequence of operations to be recorded.

The second group includes methods that allow you to determine the vector of potential or existing security events, taking into account the organization of digital segments, to restore the vector when investigating a security incident [8]. Statistical methods. Markov processes for detecting LAN security states. Viterbi algorithm for reconstructing the security event vector.

Thus, the novelty of the study lies in the use of a combination of several methods for analyzing the signs of security events with expert methods for controlling the analysis, focusing on computational methods based on Markov processes. The complex of Markov models, taking into account the formalization of typical atukayut sequences (techniques and tactics), will make it possible to predict and identify early undetected stages in the development of harmful effects on information systems.

The event vector is formed taking into account the classification of event features. All random sequences that correspond to the Markov sequence rule are considered as follows [1]:

$$P(y_1, \ldots, y_n) = P(y_1) \prod\nolimits_{i=2}^{n} p(y_i | y_{i-1}), \tag{1}$$

where $y_{i-1} is$ the indicator is the previous compromised state of a segment of a digital system with a fixed sign of a negative event.

The process of transition from the state to the following states while observing sequences based on a functional relationship is described in the form of a discrete Markov sequence. For the selected periods of fictification of events in segments of digital systems at time Δt, the probabilities of the transition are of the form [2]:

$$\begin{aligned} p_{ii}(t, t+\Delta t) &= 1 - \lambda_{ii}(t) \cdot \Delta t + 0(\Delta t), \\ p_{ij}(t, t+\Delta t) &= \lambda_{ij}(t) \cdot \Delta t + 0(\Delta t), i \neq j, \end{aligned} \tag{2}$$

where λij(t) is the intensity of the transition, which characterizes the number of transitions from state to a new state for a certain period.

The transition probabilities p_{ij} in any time period t correspond to simple linear differential equations. When constructing a vector of security events, a sequence of related states is created, which are implemented in the form of a series of events recorded in a certain period of time, marking the state of compromise on network nodes. The sequence of states (marked events) taking into account the period of fixation and the sequence of events, classified by belonging to tactics, it is formed taking into account the possible presence of other attack vectors. To do this, it is necessary to associate tactics and states that will be the vertices of the graph describing Markov processes [9].

4 Discussion

The following should be noted:

1. All the effects under consideration assume a certain initial state (a set of probabilities), which in most cases coincides with the initial stage of the vector of security events (intelligence). However, it should be noted that detection of security events can occur at any time. Accordingly, there is a need to build a vector of security events, taking into account:

   the specifics of the dynamics of negative impacts on the segments of the system;
   the time of registration of the event, compared with a certain stage of its promotion.

Such an approach will allow you to use the vector to detect the actions of the attacker, as well as determine the possible negative and hidden consequences of the previously implemented stages.

2. It should be noted that a vulnerability (more precisely, the fact of its presence in any segment of digital systems, a measure of its applicability) is exploited when one threat or multiple threats are implemented. This indicates that the sequence of security events, as well as the simultaneous use of several threats, can be determined using a vector of event features, taking into account the specifics of system segments, for which Markov chains can be used. To do this, you must first parameterize existing security events. Some network parameters of security events that indicate active exploitation of vulnerabilities can be identified as threat markers only when they are used in conjunction with other markers or under certain conditions of their (markers) manifestation. At the same time, their fixation may indicate a simple error in data transmission in digital segments. Thus, the very detection of signature markers in som cases does not guarantee that the built vector of security events corresponds to the realities of attacking influences. Moreover, the probability of incorrect vector formation increases when receiving erroneous network packets in large-scale networks, as well as when routing devices are incorrectly configured [9]. Accordingly, the vector of security events, formed taking into account the vector of vulnerabilities, the probabilistic nature of attack markers, will be more accurate, and will also help to identify possible previously unnoticed vulnerabilities. To restore previously unnoticed, but already implemented stages of the development of security events using vectors created on the basis of Markov processes, you can use the Viterbi algorithm. This will allow not only to predict attacking influences by clarifying the type of attacking influences, but also to investigate security incidents.

3. Most of the existing methods do not consider the typology of negative impacts, which affects the determination of the vector of security events, and, accordingly, the parameters of the probability of implementation and manifestation of signs of security events at various nodes in digital segments.
4. It should be noted that a big problem is the determination of the probabilities of the initial security events of segments of digital systems. Accordingly, to determine the incoming probabilities, you can use various means of registering events and their intellectual analysis, fixing a security event and using machine learning methods, neural network analysis, and other probabilistic models.

5 Conclusion

The practical significance of the research results in the field of applying a hybrid approach to the formation of a vector of security events lies in the ability to increase the level of security and stability of the segments of information systems, to quickly identify problems and anomalies. In addition to improving the overall level of infrastructure security, the proposed architectural and algorithmic solutions can be used in the design of new information infrastructures and in the management of information security of digital systems.

In the management of information security of organizations, rapid response is proposed to be carried out on the basis of a new set of system modules focused on the collection and full-scale analysis of data from system segments. At the same time, the topology of the system of digital segments of the enterprise, the value of protected information resources are taken into account.

References

1. GOST R 56546-2015 National Standard of the Russian Federation. Protection of information. Vulnerabilities of information systems. Classification of vulnerabilities of information systems. Moscow: Standartinform (2018)
2. Stevens T., Ertan A., Floyd K., Pernik P.: Cyber Threats and NATO 2030: Horizon Scanning and Analysis [Electronic resource]. https://ccdcoe.org/uploads/2020/12/Cyber-Threats-and-NATO-2030_Horizon-Scanning-and-Analysis.pdf. Accessed 14 July 2022
3. Yu, O.V., Kondratyuk, A.P.: Evaluation of information in the interests of reflexive management of competitors. Softw. Prod. Syst. **2**, 64–68 (2010)
4. Hentea, M.: Intelligent system for information security management: architecture and design issues. Issues Informing Sci. Inform. Technol. **4**, 15–20 (2007)
5. Brotby K.: Information security governance. A Practical Development and Implementation Approach, pp. 185. Wiley, Hoboken (2009)
6. Sun, L., Srivastava, R.P., Mock, T.J.: An information systems security risk assessment model under Dempster-Shafer theory of belief functions. J. Manage. Inform. Syst. **22**(4), 109–142 (2006)
7. Susanto, H., et al.: Digital ecosystem security issues for organizations and governments. Digit. Ethics Privacy (2021). https://doi.org/10.4018/978-1-7998-4570-6.ch010
8. Burgess, M., Canright, G., Engø-Monsen, K.: A graph-theoretical model of computer security. Int. J. Inf. Secur. **3**(2), 70–85 (2004). https://doi.org/10.1007/s10207-004-0044-x
9. Dewri, R., Ray, I., Poolsappasit, N., Whitley, D.: Optimal security hardening on attack tree models of networks: a cost-benefit analysis. Int. J. Inform. Secur. **11**, 167–188 (2012)

Prospects for the Development of Innovative Banking Ecosystems Based on the Digitalization of Business Processes

Ekaterina Kharlamova(✉), Artem Oleynikov, Irina Ezangina, Irina Chekhovskaya, and Tatiana Boriskina

Volgograd State Technical University, Volgograd, Russia
sikaterina@mail.ru

Abstract. In today's economy, there is a strong development of innovation ecosystems, especially in the banking sector. Banks are striving to create digital platforms and innovative solutions to provide more efficient and convenient services to their customers. Digitalization of banking processes and implementation of new technologies have become a necessity for banks to be competitive in the market. Innovative banking ecosystems offer a wide range of services, from online banking and mobile applications to the use of artificial intelligence and blockchain technologies. This research paper explores the prospects for the development of innovative banking ecosystems based on the digitalization of business processes. The paper analyses the theoretical foundations of the study of innovative banking ecosystems, analyses the key players in the market, such as VTB PJSC, SBER PJSC, Tinkoff Bank JSC, Alfa Bank JSC, Rosselkhozbank JSC and Gazprombank JSC. The analysis revealed factors that influence the formation of effective banking ecosystems, such as the use of modern information technologies, the development of digital customer service channels, the automation of business processes and the introduction of new technologies. Based on the analysis, measures were proposed and prospects for the development of innovative banking ecosystems were identified, including further digitalization of business processes, introduction and use of artificial intelligence and machine learning, and development of blockchain technologies.

Keywords: Ecosystem · Banking Ecosystems · Innovative Banking Ecosystems Digitalization · Business Processes · Competitiveness · Ecosystem Services · Information Technologies

1 Introduction

Today, an effective service ecosystem is the most important prerequisite for commercial banks and the core of their new strategies. It is through services that hundreds of millions of people become customers of the respective banks.

The influence of ecosystems is growing day by day, which has a significant effect on the transformation of the economy. The development of digital ecosystems affects

R. Polyakov (Ed.): EcoSyst 2023, LNNS 1092, pp. 221–236, 2024.
https://doi.org/10.1007/978-3-031-67354-2_24

the functioning of various socio-economic agents, creates prerequisites for a significant change in approaches and principles of management of both the national economy as a whole and industries, sectors, spheres of activity, including various business entities. "According to McKinsey forecasts, in 2025, about $60 trillion of revenue will be generated by ecosystems, and the share of corporate profits from integrated platforms in total revenue will grow from 1% in the present to 30% in 4 years' time" [1].

Large banks currently offer a wide range of financial and non-financial products that are sold through ecosystems, which limits other players' access to their distribution channels. This creates non-market competitive advantages for those organizations that have gained access to the network, strengthens the position of key players - ecosystem administrators and forms barriers to the growth of financial services providers whose access to the ecosystem has been restricted.

At the same time, the multipolar nature of ecosystems makes it difficult for industry regulators to interact with them. In addition, the ecosystem can successfully maneuver across different regulated activities, creating regulatory arbitrage. Finally, ecosystems act as a closed alternative to an open market where everyone competes with everyone else, which means that the principles of free market relations are violated.

2 Theoretical Bases for the Study of the Concept of "Ecosystem" and Preconditions of Its Emergence

Today, competition in the marketplace is increasingly less about the diversity of value propositions and more about finding new ways to co-operate and collaborate. As opportunities grow, so do the challenges that CEOs face in digital innovation ecosystems. Such ecosystems are complex, but nevertheless, they must be adapted to. To succeed, the ecosystem must be closely aligned with the core business, and ecosystem participants must move from rigid strategies to dynamic experimentation and continuous exploration of new solutions. For example, connecting large companies and users through networks, allows different values to be distributed directly between them [2].

The concept of "ecosystem" was born back in the 30 s of the XX century thanks to the works of A. Tensley. From the point of view of biological science, an "ecosystem" is a set of compactly existing living organisms in relationship with their environment, an ecosystem is in constant energy exchange with the natural environment, i.e. it is an open system. This concept was introduced directly into economic science by J. Moore, according to whose works "an ecosystem is an economic community consisting of the firm itself, its consumers, suppliers, market intermediaries, channels of goods movement, owners and other stakeholders, governmental and non-governmental organizations, and even competitors". In his report, Deloitte Touche gives his definition of business ecosystems: "Ecosystems are dynamic and co-evolving communities made up of diverse actors that create and receive new content through both interaction and competition." [3].

Ecosystems make it possible to get almost everything a consumer needs daily in one environment, one application. According to N. I. Bykanova, "the basis for the emergence of ecosystems was the changes in consumer behavioral preferences caused by the digitalization of banking activities". Indeed, banks monitor the behavior of their customers and offer them services and goods that they chose earlier [4].

One of the key prerequisites for the creation of banking ecosystems in Russia is recognized as a decline in income from traditional banking operations and the desire of banks to increase profits at the expense of non-banking income, which provoked credit institutions to refocus on the "sale" of non-banking services.

The first step towards the formation of modern banking ecosystems back in the early 21st century was the creation of insurance subsidiaries by credit organizations. At present, "banks continue to search for additional sources of income, which can be diversified by selling non-banking services» and continue to develop the spheres of consumer goods and services.

Current Russian legislation does not define the concept of "banking ecosystem". Nor has the legislator provided for the basic concept of "ecosystem". This definition is widely discussed in the literature, with researchers using the concepts of "ecosystem", 'digital ecosystem", "information system", "digital economy ecosystem", "business ecosystem"; authors often do not even distinguish between these concepts (Fig. 1). It should be agreed that the general theory of ecosystems is currently in its infancy and works that systematize approaches to the study of ecosystems have been published only in recent years. Next, let us consider the types of ecosystems and their concepts.

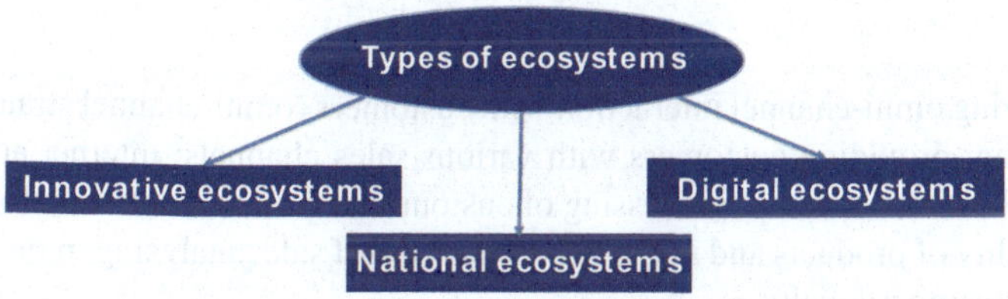

Fig. 1. Types of ecosystems [compiled by the authors]

The innovation ecosystem is a set of organizations, institutions, entrepreneurs, investors, and other actors that interact and cooperate to create and develop innovations. It provides conditions for the emergence of new ideas, their commercialization, and the diffusion and application of innovations in different areas of the economy. In an innovation ecosystem, strong linkages between actors, access to funding, scientific and technical expertise, and government support are important. In turn, a digital ecosystem is a network of interconnected digital platforms, applications, technologies, and actors that enable the creation, exchange, and use of digital information. The digital ecosystem includes various companies, organizations, public institutions, and individuals who use digital technologies to solve problems and meet needs in different areas of life. An important element in the digital ecosystem is the accessibility and usability of digital resources, as well as data protection and information security. The national ecosystem is a system of interaction and interdependence between different sectors of the national economy, government agencies, educational institutions, entrepreneurs, and other participants in the national system. The national ecosystem includes all elements that contribute to the development of a country's economy, including infrastructure, legal and political system, financial sector, scientific and technological capacity, culture, and education. The crucial aspect of the national ecosystem is to support innovation, promote entrepreneurship and create favorable conditions for the development of the country's economy [5, 6].

Meanwhile, the following components are necessary to build a digital ecosystem in commercial banks (see Fig. 2):

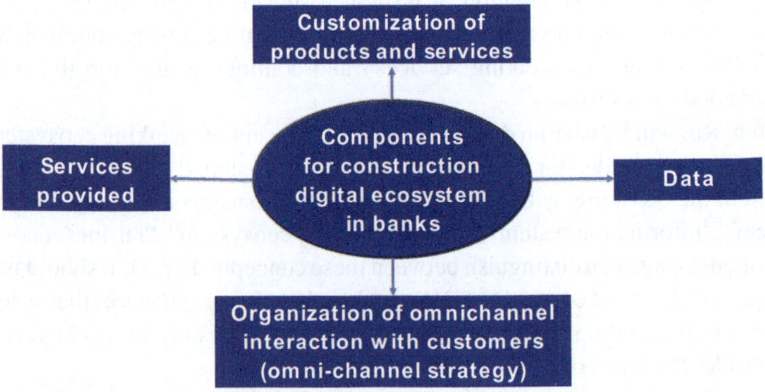

Fig. 2. Components for building a banking ecosystem [compiled by the authors]

Organizing omni-channel interaction with customers (omni-channel strategy) means the following: providing customers with various sales channels: internet applications, mobile applications.; prompt processing of customer enquiries; accurate information on the availability of products and services at each point of sale; analysing, monitoring, and managing customer loyalty.

Services provided: this component should fulfil attributes such as flexibility and adaptability. Banking services should become an everyday norm of customers' lives.

Data: ensuring that it is processed quickly and intelligently in the customer relationship process.

Customization of products and services: the development of the product required by the customer should be based on constant interactions with the customer to improve the software (Service agility) [7].

Thus, we can say that this format of doing business in the banking sector provides several advantages to its participants, as it brings together many companies of different profiles, satisfies a wide range of needs, creates a complete environment for consumers and increases the income of companies. Ecosystems integrate mechanisms that provide a simple transition from service to service, increasing the comfort level of users. Companies that create such an innovative environment stop focusing on the product and concentrate on building a unified customer journey, which significantly increases their chances of attracting and retaining the maximum number of customers.

3 Analysis of the Ecosystem Structure of the Leading Russian Banks

Russian banking ecosystems were unknown abroad just a few years ago, but now the cases of Sber, Tinkoff Bank and Alfa Bank are being studied all over the world. Other players - VTB, Gazprombank, and Rosselkhozbank - are gradually gaining momentum. At the same time, the ecosystems of Russian banks must compete not only among themselves, but also with high-tech corporations such as Yandex, MTS and VK. In this regard, it is extremely important to study the approaches and experience of ecosystems functioning, including for determining the vector of economic sector development in Russia [8].

Modern Russia is pursuing economic growth through the introduction of breakthrough innovations in the national infrastructure, which subsequently enables the country to become one of the world's digital leaders. By many parameters of digitalization, Russia is among the leading countries - innovations and online services are demanded by the population, and there are technological and infrastructural conditions for their accelerated development.

In the Russian banking sector, we can distinguish the "Big Six" banks, including Sberbank, Tinkoff Bank, VTB, Gazprombank, Rosselkhozbank and Alfa-Bank (Fig. 3).

Fig. 3. Leading Russian banks with their own ecosystems [compiled by the authors]

Large banks actively compete for the mass client, thus determining the situation in the banking industry. Leaders of the banking sector in the competitive environment build platform and ecosystem business models, thus taking less strong players in the market out of the game [9].

One of the features of the Russian banking services market is the significant role of the financial sector in the formation of ecosystems - large banks are actively moving in this direction. At the same time, technology companies are building financial services into their ecosystem product line [4]. Due to their scale, digital ecosystems have a significant impact not only on individual industries, but also on the country's economy. Modern ecosystems accelerate the development and implementation of new technologies in economic processes at a significant scale for the entire market.

The development of platform services has led to a significant improvement in the quality of life of consumers: services have become more accessible, goods (due to competition between suppliers) have become cheaper, and choice has become wider.

Now, 6 banks in Russia are developing ecosystems: Sber, VTB, Tinkoff Bank, Alfa-Bank, Rosselkhozbank and Gazprombank. Let's review and analyze each of the existing banking ecosystems in more detail.

Over the past few years, SBER has been actively investing in the development of its technological capabilities, with the company focusing on becoming not only a bank, but also a full-fledged technology company. For the same purpose, in 2020 the company was rebranded, and the word "bank" was removed from its name.

The main goal of the new Strategy of PJSC Sber until 2024 is to build an integrated ecosystem. The new Strategy pays special attention to building a better seamless customer experience and creating a unique offer for the customer. PJSC Sber does not look at business areas as separate assets, but as an integrated business model - the Ecosystem. The goal is to become a trusted assistant and navigator in a changing world: for the individual, business, and the state.

The ecosystem of PJSC SBER began its development in 2016, in 2017 the company started to expand its non-banking product line by entering into partnership agreements with the largest IT corporations in Russia. Currently, the ecosystem of PJSC SBER includes more than 50 companies and 120 services in many areas: e-commerce, ready-to-eat food delivery, taxis, carsharing, media and entertainment, mobile services, healthcare, real estate transactions, etc. The audience of non-financial services is more than 60 million people per month. PJSC Sber also has a unified login system SberID, which gives access to 128 different services, is used by more than 13 million customers, and the number of subscribers to the SberPrime service package is close to the number of subscribers. SberPrime has approached one million customers [18].

The following services are already offered as part of the SberX ecosystem:

SberBox – multimedia set-top box,

SberPortal – a multimedia smart screen,

SberSound – a music service,

SberDisk – cloud storage,

SberPay – contactless payment service,

SberClass – educational platform,

SberBoxPrime – access to Okko online cinema,

DomClick – services in property transactions,

SberHealth – online consultation with a doctor,

SberEPharmacy – allows you to order medicines, cosmetics, or pet products with delivery to your home or the nearest pharmacy,

SberMarket – service for delivery of groceries from shops,

Samokat – grocery delivery,

Sber Insurance – mobile operator,

SberMobile – mobile operator,

SberSpasibo – loyalty programme,

SberInvestments – trading platform with investments,

SberBusiness – a comprehensive solution for company development,

SberLogistics – mail services,

Work.ru – job search service,

SberAuto – a service for selling and buying cars.

PJSC Sber actively continues to cover different areas, up to the creation of cartoons, virtual set-top boxes for TV or smart displays for operative communication with the bank. An app shop is expected to be launched. Figure 4 shows the scheme of PJSC Sber's ecosystem.

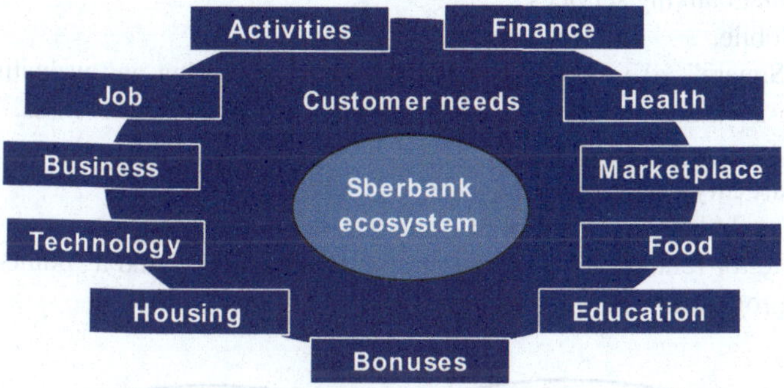

Fig. 4. Scheme of PJSC Sber's ecosystem [compiled by the authors]

Tinkoff Bank's ecosystem: the Bank provides a range of financial services for individuals and businesses. Tinkoff Bank pays special attention to the development of lifestyle banking: the ecosystem enables customers to analyze and plan personal spending, invest savings, receive bonuses under loyalty programmers, book trips, buy cinema tickets, and book tables in restaurants (Fig. 5).

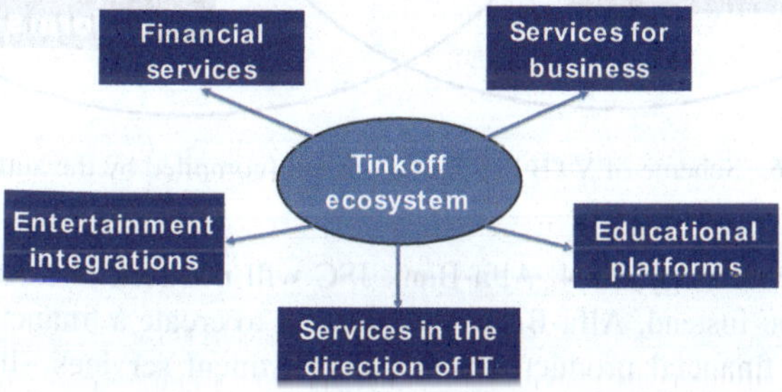

Fig. 5. Scheme of Tinkoff Bank JSC's ecosystem [compiled by the authors]

JSC Tinkoff Bank has the following services in its ecosystem:
Tinkoff Bank - traditional banking services,
Tinkoff Business - a platform for small and medium-sized businesses,
Tinkoff Investments,
Tinkoff Insurance,
Tinkoff Mobile - mobile operator services,
Tinkoff Call Defender - an anti-fraud platform to protect against phone scammers,
Tinkoff Education - educational courses,
Tinkoff Magazine,

Tinkoff Travel.

VTB PJSC has been developing its digital ecosystem since 2019 by connecting new projects to it. In the future, it is planned to create a full-fledged superapp with access to various services.

As part of the ecosystem, VTB PJSC already offers (see Fig. 6):

Traditional banking services,

VTB Mobile,

"Meter Square" - a service for searching, repairing, checking, and evaluating housing,

Unicom-24 financial services marketplace, through which cards, loans and microloans from various lenders are issued, and much more,

My Smart City online service,

Educational platforms (School of Analytics, School of Investor),

A service for renting accommodation, online accounting, and a "points exchange" for loyalty programmers are planned.

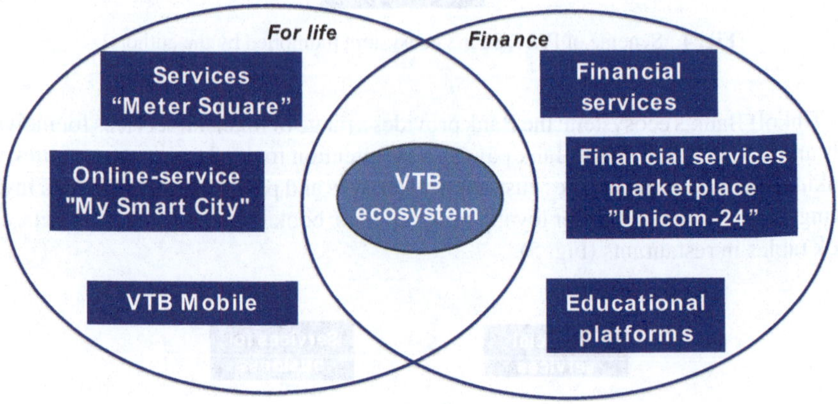

Fig. 6. Scheme of VTB PJSC's ecosystem [compiled by the authors]

According to Strategy 2024, Alfa-Bank JSC will not seek to develop its business in non-core areas. Instead, Alfa-Bank JSC is going to create a financial super-service, combining only financial products: banking, investment services, digital mortgages, savings accounts, mobile cash desks and other services. Everything else that is not related to finance will be developed by Alfa-Bank's partners in their own sales channels and interfaces [10].

Services of the Alfa-Bank JSC ecosystem:

Traditional financial services.

Alfa Investments.

Alfa Factory, Alfa Campus - education.

Alfa-Bank for Business ("Supervisory Authority Check", "Employee Check", "Risk Indicator").

Partner services (blockchain platform for settlements in the construction industry (partner Gaskar Group); a service for buying tickets and booking hotels (partner One two trip).

The scheme of Alfa-Bank's financial super-service is as follows (see Fig. 7).

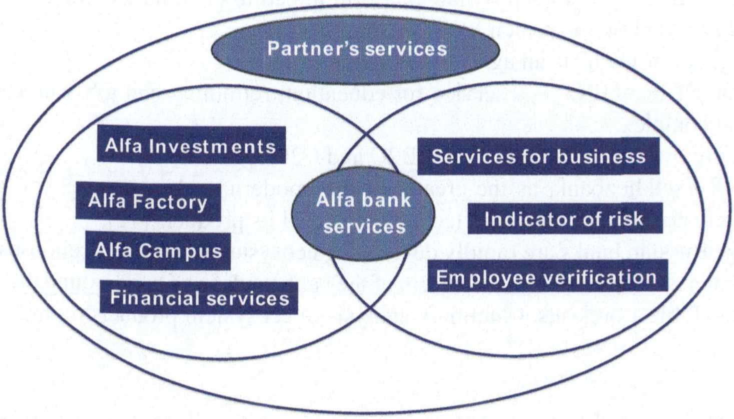

Fig. 7. Alfa-Bank JSC's Financial Super Service Scheme [compiled by the authors]

The management of JSC Gazprombank believes that for many customers, the bank is a business partner where they can receive both core banking and non-banking services. Understanding the customers' needs, the team decided to create the JSC Gazprombank Group - a developed network of co-operations of different companies. It is ready to cover almost all customer tasks: from searching for business partners to concluding, paying for, and supporting contracts with contractors and customers.

The main goal is to unite all JSC Gazprombank Group's offerings into a single tool based on a common digital ecosystem where customers have no boundaries between remote service providers [11].

JSC Gazprombank's ecosystem: Traditional financial services, GPB Mobile, DevSchool – an educational platform, Gorodpay – a mobile application with a wide range of opportunities for public transport passengers,

Privilege Programme "Sport" – a subscription for 110 points every month for fitness classes and visits to swimming pools at partner wellness centers.

Gazprombank's ecosystem is currently developing and expanding. The Bank aims to develop services that are based on customers' needs and are designed to simplify their work.

Rosselkhozbank's ecosystem is designed for micro, small and medium-sized businesses, as well as for people who just want to start their own business in the agricultural sector. Its main objective is to support existing farmers and create conditions for increasing their number.

"Own" ecosystem is currently represented by seven sites:

"Own Farming" - designed to ensure that farmers are aware of digital innovation solutions available on the market and can implement them.

"Own Native" – is an extensive offer of farm products. On this website, every farmer has the right to present, free of charge, full information about his farm, about the products he produces.

"Own Dwelling" – is a set of mortgage programmers - primarily, the programmer of preferential rural mortgages and banking offers from JSC Rosselkhozbank.

"Own Selo" – a completely new platform aimed at creating a comfortable life in rural areas, based on the principles of partner services.

"Own out of town" is an agro-tour booking service.

"Yourself in AGRO" – a service for education, retraining and job search in agro-industrial complex.

"All my own" – a marketplace for B2C and C2C segments.

JSC Rosselkhozbank, as the creator of this modern ecosystem, seeks to unite all those interested in agriculture, its technologies, and its products [12].

Large Russian banks are rapidly developing ecosystems, each bank chooses its own development strategy, but any ecosystem of the bank tends to be built around the needs of the client. Table 1 presents a summary analysis of ecosystem products of large Russian banks.

Table 1. Products and services included in the ecosystems of major Russian banks [compiled by the authors]

Service	SBER	Tinkoff Bank	VTB	Alfa bank	Gazprombank	Rosselkhozbank
E-commerce	+	–	–	–	–	+
Real estate	+	+	+	–	–	+
B2B products	+	+	–	+	–	+
Investments	+	+	+	+	+	–
Insurance	+	+	+	–	–	–
Job Search	+	–	–	–	–	+
Travelling	+	+	+	+	+	+
Mobile communication	+	+	+	–	+	–
Financial services	+	+	+	–	–	–
Education	+	+	+	+	+	+
Food delivery	+	–	–	–	–	–
Loyalty	+	+	+	+	+	+

Having analyzed the data in Table 1, we see that modern financial ecosystems offer clients and users an extensive range of services. The undisputed leaders in building the ecosystem are PJSC Sber and JSC Tinkoff Bank. VTB PJSC, Alfa-Bank JSC and Gazprombank JSC are at the stage of developing their ecosystem products. Rosselkhozbank is building one of the first digital ecosystems for agro-industrial and farming enterprises.

4 Factors Influencing the Formation of Effective Banking Ecosystems

To achieve successful development and existence of a bank, as well as growth of profits and other financial indicators, a modern bank should strive for the following trends:

1. Full digitalization. The bank should be able to process many transactions, because due to digitalization all transactions, even the smallest in size, go cashless, and a huge flow of transactions is formed, which today not all banks are ready to process in real time [13].
2. Formation of a digital portrait of each of its clients. The bank should understand all the needs of the client: from how often he orders products, how often he uses a taxi. Based on the collected information, offer the client an appropriate banking product. For someone it will be cash payments with a minimum commission, and for someone it will be lending with special conditions.
3. To strive for adaptability, to adjust flexibly and quickly to the needs of the client with minimal human involvement. Today the client needs a bank that will solve his problems for him.
4. Satisfaction of all possible needs of the client. The bank should offer financial and non-financial products to the client. And if financial services used to be the main service, at the stage of development of banking ecosystems, they would only accompany and support the provision of non-financial services.
5. Maintenance of all ecosystem processes round the clock.
6. High level of banking security system. Reliable protection of confidential data and documents [5, 13].

Russian customers today prefer those products that correlate with their rhythm of life and help to solve everyday tasks easier and faster. For example, the choice of a bank can be conditioned by a person's profession, income level and even interests. The success of bank ecosystems depends on several factors (see Fig. 8).

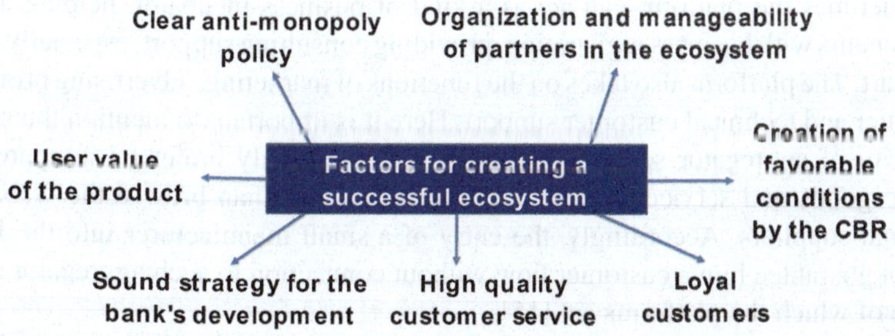

Fig. 8. Key factors for building an efficient banking ecosystem [compiled by the authors]

The identified factors of ecosystem development in the banking sector will allow the formation of an adequate mechanism for the implementation of innovation processes, increase the availability and quality of technology, products, and services, as well as increase the level of bank competitiveness.

5 Prospects for the Development of Modern Innovative Banking Ecosystems

The development of ecosystems opens the following prospects for banks:

1. The opportunity to accelerate customer acquisition and revenue growth through cross-selling. This is especially relevant for large banks, whose growth prospects in traditional segments are limited due to their high share.
2. The ecosystem can be a prerequisite to compete effectively with technology companies that are part of the financial services segment.
3. A diversified ecosystem adapted to the daily needs of specific users can become an effective tool for a bank to increase customer loyalty and retention [14].

The development of platform services has led to a significant improvement in the quality of life of consumers: services have become more accessible, goods (due to competition between suppliers) - cheaper, and the choice - wider.

The transition from the initial "noticeboard" format to a centralized platform model with the platform setting rules for providers has increased the reliability of services and consumer protection. Given that the basis of network effects is mass (and in bilateral markets, mass of both sellers and consumers), it is not surprising that the main goal of platforms, especially in the first stages of development, is to attract customers.

When the ecosystem receives data on customer actions and purchases, it can generate targeted offers that are tailored to the customer - for example, if a customer searches for information on painting courses, they will also be offered the opportunity to purchase art supplies online. This can significantly save the customer time spent searching for and purchasing the goods and services they need.

There are also benefits for suppliers from the development of platform business models. These are new markets that they themselves may never have reached due to geographical barriers. This has the greatest effect for small and medium-sized businesses with limited resources to expand their network of outlets. Typically, such a business is physically present within a small region.

Sometimes the platform can act as a kind of business incubator, helping aspiring entrepreneurs with business registration, providing consulting support, especially needed at the start. The platform also takes on the functions of marketing, advertising promotion, call-center and technical customer support. Here it is important to mention the growing popularity of aggregator services - consumers increasingly prefer to compare offers of basic goods and services in a single interface rather than browse the websites of individual suppliers. Accordingly, the entry of a small manufacturer into the Internet does not guarantee him a customer flow without connection to such aggregator sites, in the role of which the platforms act [15].

The disadvantages of ecosystems for economic development include the planes of market constraints and market oligopolist. Excessive power of the largest participants can manifest itself in different sectors, not only in the financial market, but also in a few segments where they sell goods and services. An oligopoly will be able to close the entrance or set high barriers for new entrants to the market, resulting in lower differentiation and quality of services, and higher prices. Increased competition can be especially sensitive for small companies that do not have sufficient financial resources to invest in the digital transformation of providing a high level of service, the same as the largest players in the market [5].

From this analysis, we can identify the main factors of ecosystem impact on the country's economy (see Table 2).

Table 2. Positive and negative impacts of ecosystems on a country's economy [compiled by the authors]

№	Exposure factor	Impact on the economy
Positive factors		
1	Increase in the bank's capital and assets	Large banks can participate in state investment programmers aimed at developing the country's economy
2	Emergence of new markets for small and medium-sized businesses	Opportunity to develop small companies with limited resources, removing geographical constraints on the development of their business
3	Accessibility of services to the consumer	Improving the quality of life of consumers
Negative factors		
1	Monopolization of the banking sector by large players	Reduction of banking competition Establishment of high barriers in the market for new entrants, resulting in lower differentiation and quality of services, and higher prices
2	Increased risks of leakage and misuse of personal information for commercial interests	Decrease in the level of public confidence in the country's financial sector
3	Challenges of Legislative and Regulatory Regulation of Ecosystems	Ecosystems could take over some of the market and become the analogue of a regulator within its perimeter

Based on the analysis of banking sector ecosystems, the authors propose the following measures for the development of innovative banking ecosystems:

1. Introduction of intelligent chatbots in most of the banking operations performed by customers. This technology will help in fully automating customer interactions. This will facilitate instant transactions, the ability to offer customized products and services on an ongoing basis and completely abandon the street banking scenario.
2. The development of digital self-service channels in mobile and online banking, the introduction of this technology increases the operational efficiency of the banking sector and provides superior customer service.
3. Transition from face-to-face transactions to fully digital services. Banks can offer enhanced targeted services on a range of new devices that duplicate the functionality of smartphones but are more convenient and portable.

4. Increased partnerships between banks and established financial-technical institutions. Financial tech companies can provide established and proven technology banking services to banks. Banks can offer access to a much broader customer base and influence the image of security and stability of financial-technical companies.
5. Personalization of the customer experience. The interaction between bank and consumer should be based on knowledge of the customer's needs, as well as past experiences and requirements. Personalization for the customer means that they will build a deeper relationship with their bank and be less likely to look at prices in a highly competitive market. To deliver a fully personalized experience, a bank needs comprehensive and relevant data to which it can apply the power of artificial intelligence and machine learning.
6. The development and refinement of new technologies such as artificial intelligence (AI), blockchain and the Internet of Things (IoT) will increasingly impact the banking industry.
7. The introduction and development of remote biometric identification mechanism, digital citizen profile infrastructure, fast payment system, marketplace platform will contribute to the security, availability of financial and other services in digital form, and the development of competition [11, 16, 17].

6 Conclusions

Thus, ecosystems bring innovation to the economy and provide a new quality of life for consumers. Banking ecosystems provide new opportunities for suppliers of goods and services. Undoubtedly, the development of ecosystems in the banking services markets is a powerful boost for the economy of a country and in general. However, the rapid and uncontrolled development of ecosystems has several negative factors, but with adequate regulatory and supervisory measures, ecosystems will be a powerful driver for the development of the country's economy. Also, we can say that today banks are not only about financial services, but also about creating new holistic ecosystems that create a comfortable environment around one person (or business), surrounding it with care and necessary services.

The analysis of innovative banking ecosystems confirms the importance of digitalization in the development of innovative banking ecosystems. The analysis conducted on the example of such large banks as VTB PJSC, SBER PJSC, Tinkoff Bank JSC, Alfa-Bank JSC, Rosselkhozbank JSC and Gazprombank JSC allows us to identify the key factors influencing the formation of effective banking ecosystems. Based on the results of the analysis of banking ecosystems, we have identified areas of action and prospects for the development of innovative banking ecosystems: the development of digital self-service channels in mobile and online banks, new technologies such as artificial intelligence (AI), blockchain, and the introduction and development of a remote biometric identification mechanism.

Current trends in the development of the banking sector indicate that in the near future banks will no longer be just financial institutions but will transform into ecosystems that accumulate not only banking services, but also the services of partner companies. Having access to a huge audience of their customers, they can typify them into certain segments and offer them solutions that would cover their needs. Everything is aimed at ensuring that the bank's client receives the widest possible range of services that would allow them to solve the problems of one or another part of their daily life.

References

1. How do companies create value from digital ecosystems. https://www.mckinsey.com/~/media/McKinsey/Business%20Functions/McKinsey%20Digital/Our%20Insights/How%20do%20companies%20create%20value%20from%20digital%20ecosystems/How-do-companies-create-value-from-digital-ecosystems-vF.pdf. Accessed 19 January 2024
2. A look inside the new trends in business. https://www.fastcompany.com/3045104/a-look-inside-the-new-trends-in-business. Accessed 14 January 2024
3. Moore, J.F.: Predators and prey: a new ecology of competition. Harv. Bus. Rev. **71**(3), 7–86 (1993). PMID: 10126156, https://pubmed.ncbi.nlm.nih.gov/10126156/. Accessed 18 January 2024
4. Bykanova, N.I., Solovey, Y.A., Gordya, D.V.: Formation of ecosystems of banks in the conditions of digitalization of the banking space. Finance State Enterprises **1**(47), 94–95 (2020)
5. Carayannis, E.G., Grigoroudis, E., Campbell, D.F., Meissner, D., Stamati, D.: The ecosystem as helix: an exploratory theory building study of regional co-operative entrepreneurial ecosystems as Quadruple/Quintuple helix innovation models. R&D Manage. **48**(1), 148–162 (2018)
6. Roles and objectives of modern central banks. https://analystprep.com/cfa-level-1-exam/economics/describe-roles-objectives-central-banks. Accessed 15 January 2024
7. Nurmukhametov R.K., Voskresenskaya L.N., Myasnikova E.B.: Banking ecosystems in Russia: essence, types, regulation. Financial Markets Banks **8**(1), (2021). cyberleninka.ru/article/n/bankovskie-ekosistemy-v-rossii-suschnost-vidy-regulirovanie. Accessed 20 January 2024
8. Bolsun, V.S.: Features of digital technologies implementation in the banking sector of Russia. In: Actual problems and prospects of development of innovative economy and management: a collection of scientific articles, pp. 166–171 (2019)
9. Index of penetration of fintech services 2019. https://assets.ey.com/content/dam/ey-sites/ey-com/ru_ru/news/2019/11/ey-fai-2019-rus.pdf. Accessed 18 January 2024
10. Kleiner, G.B.: Development of ecosystems in the financial sector of Russia. Ecosyst. Approach Bus. Transform. Manage. Theory Pract. **4**(11), 4–7 (2021)
11. Digitalization of banking products as the basis for the formation of the bank ecosystem. www.ibispb.ru/docs/science/dissertation-council/d-999-076-02/bychkova/BychkovaII.pdf. Accessed 14 January 2024
12. Rubinstein, T.B.: Development of the banking system and innovative banking products. Bank. Serv. **3**, 11–14 (2018)
13. Kapoor, R.: Ecosystems: broadening the locus of value creation. J. Organ. Des. **7**(1), 12–17 (2018)
14. How banks are using ecosystems to drive growth and profits. https://www.ey.com/en_us/banking-capital-markets/how-banks-are-using-ecosystems-to-drive-growth-and-profits. Accessed 15 January 2024
15. Global FinTech Adoption Index. https://www.ey.com/en_gl/ey-global-fintech-adoption-index. Accessed 19 January 2024
16. Frost, J., Gambacorta, L., Huang, Y., Song Shin, H., Zbinden, P.: BigTech and the changing structure of financial intermediation. Monetary and Economic Department (2019)

17. Beyond the traditional bank: the race for ecosystems!. https://skaleet.com/en/blog/beyond-the-traditional-bank-the-race-for-ecosystems. Accessed 16 January 2024
18. Gimpel, H., Rau, D., Roglinger, M.: Understanding FinTech start-ups - a taxonomy of consumer-oriented service offerings. Electron. Mark. **28**(3), 245–264 (2018)

Sustainable Mobility. Electric Vehicles as an Environmentally Friendly Mode of Transportation

Sergey Muzalev(✉), Irina Demina, Tatiana Izzuka, and Natalia Zvonareva

Financial University Under the Government of the Russian Federation, Moscow, Russia
svmuzalyov@fa.ru

Abstract. In the modern world, within the context of current global climate change, sustainable development is not only necessary but also the only viable path for societal progress. Given the rapid growth of modern cities, where the number of automobiles on the streets increases every year, it becomes evident that urban planning needs to be reevaluated in alignment with the principles of sustainable mobility. In large cities, automobile exhaust emissions release hundreds of thousands of cubic meters of pollutants and greenhouse gases annually, causing significant harm to the environment and public health. In this regard, the development of electric transportation, particularly electric vehicles, represents an environmentally friendly and efficient alternative to contemporary transportation. This article examines the current situation of automotive transportation in Russia and the level of damage caused by emissions from internal combustion engine vehicles. Additionally, the article explores the fundamental aspects of sustainable development and sustainable mobility concepts, as well as measures to implement sustainable mobility in modern cities. It analyzes the current state of the electric vehicle industry in Russia and globally, and discusses measures to intensify the development of this industry in our country.

Keywords: Sustainable Development · Sustainable Mobility · Transportation Electrification · Electric Transportation · Electric Vehicles · Urban Planning · Automotive Transportation

1 Introduction

One of the evident consequences of the intensive scientific and technological progress of the 20th and 21st centuries has been a significant increase in the urban population. The high degree of urbanization in modern countries is a phenomenon characterizing the current stage of human development. In the overwhelming majority of states, there is a trend of rural population migration into small and medium-sized cities, as well as into large metropolises, which, in turn, are growing in size each year, evolving into vast urban agglomerations. For instance, over the past 20 years, the population of Moscow has increased by 25%, and its area has expanded by 140% (see Fig. 1).

R. Polyakov (Ed.): EcoSyst 2023, LNNS 1092, pp. 237–247, 2024.
https://doi.org/10.1007/978-3-031-67354-2_25

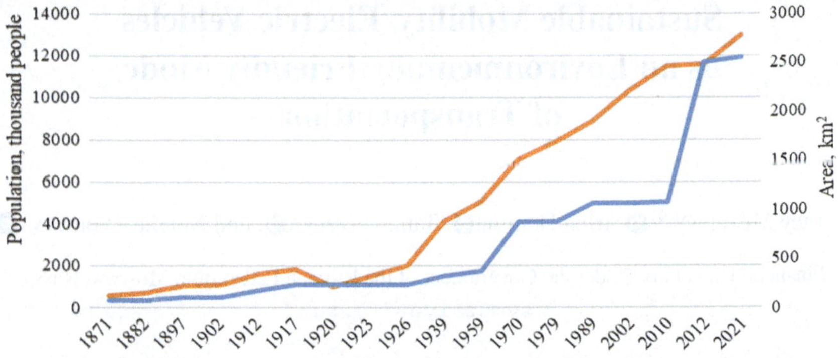

Fig. 1. The Population and Area Dynamics of Moscow City from 1871 to 2021 [1].

The obvious consequence of such rapid urban population growth is the deterioration of the environmental situation in major metropolises. Several objective reasons for the emergence of environmental issues can be identified:

- High population density in limited urban areas: In cities, the population is densely concentrated within a limited territory. Given the high population growth rates, this leads to the intensification of urban development with residential and commercial structures, the destruction of natural areas and green spaces, fertile soils, and entire ecosystems.
- Concentration of industrial facilities in cities: The growth of the urban population necessitates the provision of employment opportunities, uninterrupted functioning of municipal services, and the production of various consumer goods. Consequently, large cities establish industrial zones and sectors where industrial enterprises from various sectors of activity are concentrated over vast areas, ranging from small factories to large metallurgical complexes and nuclear power plants. These industrial facilities, to varying degrees, contribute to the deterioration of the environmental situation in the city through emissions of pollutants into the atmosphere, soil, water, and more.
- The dense urban road network: Cities often serve as hubs of road networks since automobiles play a crucial role in urban mobility and the lives of urban populations. The increasing number of private automobiles and public transport vehicles on the roads can lead to elevated levels of vehicular emissions (nitrogen oxides, hydrocarbons, and other pollutants) and traffic congestion, thereby deteriorating air quality and contributing to adverse climate changes.

The main part of harmful emissions comprises emissions from industrial stationary facilities and mobile vehicles. According to data from the Federal State Statistics Service, in 2021, approximately 22.3 million tons of atmospheric pollutants were produced in the Russian Federation [2]. Of these emissions, 77% were generated by industrial facilities, while 33% were from transportation vehicles. When calculated per capita, these figures amount to 118 kg per person and 35 kg per person per year, respectively. Despite the fact that only a third of emissions in the entire Russian Federation are attributed to automobile transport, in major cities, this indicator is typically significantly higher. For example, in

2021, approximately 387.7 thousand tons of pollutants were released into the atmosphere in Moscow, with approximately 84% attributed to motor vehicles, accounting for 325.7 thousand tons in absolute terms or 24.9 kg per urban inhabitant [3]. When calculated per unit of urban area, this figure reached approximately 130 tons of emissions per square kilometer in 2021. Clearly, such values of indicators highlight the issue of automobile emissions in cities as a major challenge on the urban environmental agenda.

Automobile emissions have an extremely detrimental impact on the environment and public health. These emissions encompass various chemical substances, such as atomic oxygen, nitrogen, water vapor, carbon monoxide, nitrogen oxides, hydrocarbons, sulfur oxides, soot, and benzo(a)pyrene (see Fig. 2).

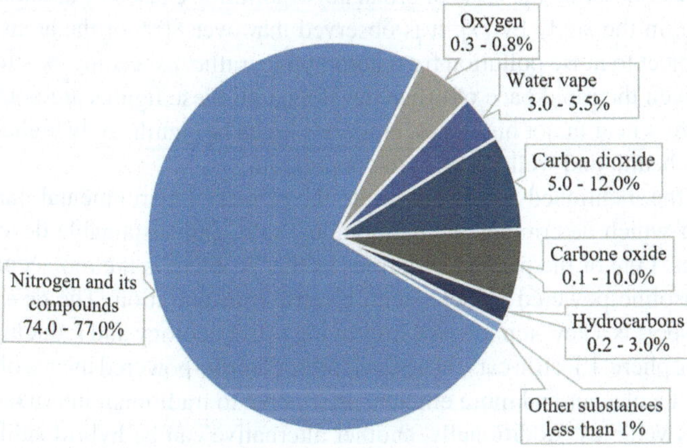

Fig. 2. The composition of harmful substances in the exhaust gases of internal combustion engines [4].

As mentioned earlier, every vehicle equipped with an internal combustion engine (ICE) is a source of a vast amount of harmful substances that pollute the urban atmosphere. For instance, an average car "consumes 2 tons of gasoline per year and emits 20–25 thousand cubic meters of combustion products into the air, containing 700 kg of carbon monoxide, 40 kg of nitrogen oxides, 230 kg of hydrocarbons, and 2–5 kg of particulate matter" [5].

The release of these chemical substances into the atmosphere leads to a range of negative consequences for the urban ecosystem. Foremost, automobile emissions are the primary source of greenhouse gases, contributing to the acceleration of global warming, the most significant climate issue of the modern world. Carbon oxides, which can make up to 22% of exhaust gases, are the main agents of the greenhouse effect, and considering the enormous volumes of annual emissions, they make a significant contribution to climate change.

The emission of these chemical substances into the atmosphere leads to a range of adverse consequences for the city's ecosystem. Primarily, automobile emissions serve as the primary source of greenhouse gases, contributing to the acceleration of global warming – the most significant climate issue of the modern world. Carbon oxides, which can constitute up to 22% of exhaust gases, are the primary agents of the greenhouse effect, and considering the vast volumes of annual emissions, they make a significant contribution to climate change.

Furthermore, nitrogen oxides and hydrocarbons emitted by automobiles are the primary drivers of smog formation and atmospheric pollution in cities. These harmful substances can lead to various health problems among the urban population, including respiratory diseases, allergies, and cardiovascular ailments. In addition to toxic compounds, exhaust gases also contain a carcinogen, specifically benzo(a)pyrene, which is a first-class hazardous chemical compound causing oncological diseases, mutagenic, embryotoxic, and hematotoxic effects, capable of entering the body through the skin, respiratory organs, gastrointestinal tract, and placenta [6].

It's also worth noting the issue of urban noise pollution caused by automotive traffic. For example, in the study by [7], it is observed that over 80% of the areas in modern cities are subject to noise pollution from automotive traffic, exceeding existing levels by 5–30 dB. Given the rapid pace of urban development, these figures are not limiting; in residential areas near major highways, noise levels can be significantly higher, adversely affecting the health and well-being of the population.

Clearly, the mentioned issues constitute the primary environmental challenge, the resolution of which determines the harmonious, safe, and sustainable development of modern cities. One of the primary solutions to this problem is the replacement of conventional gasoline-powered vehicles with electric transportation. The development of electric transport is a key step towards improving urban ecology and reducing emissions into the atmosphere. Electric cars, buses, and other electric-powered means of transportation represent a cleaner and more efficient alternative to traditional internal combustion engine (ICE) vehicles. Additionally, another alternative can be hybrid vehicles, which combine both an electric powertrain and an ICE, significantly reducing emissions of harmful substances into the atmosphere.

The aim of this research is to analyze the development of electric and hybrid vehicles, public transportation, and the infrastructure supporting eco-friendly modes of transportation.

2 Methods and Materials

Sustainable development is a globally recognized concept of societal development, first formulated in the Report prepared by the Brundtland Commission of the United Nations as "a process that meets the needs of the present without compromising the ability of future generations to meet their own needs" [8]. In its initial interpretations, this concept encompassed the harmonious development of society without harm to the environment, but over time, sustainable development has incorporated principles addressing poverty, discrimination, inequality, hunger, and so forth. In 1998, R. Costanza and K. Folke formulated three main vectors of sustainable development:

- "Maintaining a sustainable scale of the economy that does not harm the environment.
- Achieving fair distribution of natural and societal resources among social groups, different generations, as well as between humans and other biological species.

- Achieving efficient distribution of natural resources over time, taking into account their reserves and the capacity of the environment" [9].

In its modern form, the concept of sustainable development was presented at the United Nations General Assembly Summit in 2015 in the "Agenda for Sustainable Development by 2030", which adopted 17 global sustainable development goals [10].

Sustainable mobility is a concept of sustainable development for urban transportation networks, "focused on the efficient, safe, secure, affordable, and environmentally friendly movement of people and goods through internal transportation means" [11].

In conducting this research, the following sources were used as primary data: materials from the Federal State Statistics Service, the International Energy Agency, materials and resolutions from the United Nations, such as "Agenda for the 21st Century" (1992), "Habitat Agenda" (1996), the Report "Global Green New Deal" (2008), Resolution "The Future We Want" (2012), Resolution "Transforming Our World', "Agenda for Sustainable Development by 2030" (2015), "The New Urban Agenda" (2016), as well as scientific works in the field of sustainable development and sustainable mobility. The achievement of the objectives of this research was carried out using the method of content analysis of scientific publications and reports, followed by their examination, evaluation, and data systematization.

3 Discussion

The analysis of the development of electric transport should begin with fundamental concepts such as sustainable development and sustainable mobility.

The concept of sustainable development has widely spread since the 1980s as a unified strategy for reducing inequality, preserving ecosystems, and promoting the harmonious development of humanity. This strategy was solidified in the "Agenda for the 21st Century" adopted at the United Nations Conference in Rio de Janeiro in 1992 [11]. During that conference, the main goals of the strategy were formulated, namely the 17 global sustainable development goals (Fig. 3). In 2015, at the United Nations Summit, 193 member states of the UN committed to achieving these goals by 2030 [12].

The key premise behind the creation of this concept was the increasing year-by-year negative anthropogenic impact on the Earth's ecosystem due to the rapid pace of scientific and technological progress, leading to subsequent global climate changes. In accordance with this, the primary goal of sustainable development can be defined as the minimization of factors affecting the environment while maintaining the pace of technological advancement.

Speaking about the concept of sustainable mobility, it is worth noting that this concept is based on the same principles as the main concept of sustainable development discussed above. Sustainable mobility in cities implies a vector of urban transport system development in which the most advanced technologies are applied, safety and accessibility of transportation are ensured for all segments of the population, transportation chains are optimized, and the negative impact on the environment and the health of the urban population is minimized.

Fig. 3. 17 Global Sustainable Development Goals [13].

The concept of sustainable mobility is based on the following fundamental principles:

- Environmental sustainability: reducing the negative impact on the environment, particularly lowering emissions of exhaust gases through the implementation and development of electric transport, enhancing public transportation, urban infrastructure, bicycle lanes, and pedestrian pathways.
- Social inclusivity: ensuring equal access to the transportation system for all segments of the population, including individuals with limited mobility and those with low income.
- Economic efficiency: optimizing resource utilization and infrastructure to reduce operational expenses and enhance the efficiency of the transportation network through smart traffic management, integrating various modes of transportation, and improving routing.
- Urban planning: shaping the city's urban development policies using sustainable mobility principles. For instance, creating compact urban structures with well-placed points of attraction for city dwellers will make public transportation and bicycles more preferable than private automobiles.
- Innovation and technology: building a smart and sustainable urban transportation network is impossible without utilizing modern technologies such as advanced electric vehicles, charging stations, car-sharing/ride-sharing apps, and more.

A vast global expert community is actively engaged in the development and elaboration of ideas within this concept. Their proposals for achieving sustainability in transportation systems have been documented in various reports and resolutions, such as "Agenda 21" (1992), "Habitat Agenda" (1996), the report "Global Green New Deal" (2008), the Resolution "The Future We Want" (2012), the Resolution "Transforming Our World", the "2030 Agenda for Sustainable Development" (2015), the "New Urban Agenda" (2016), and so forth. The author of this study [14] has briefly summarized these proposals (see Fig. 4).

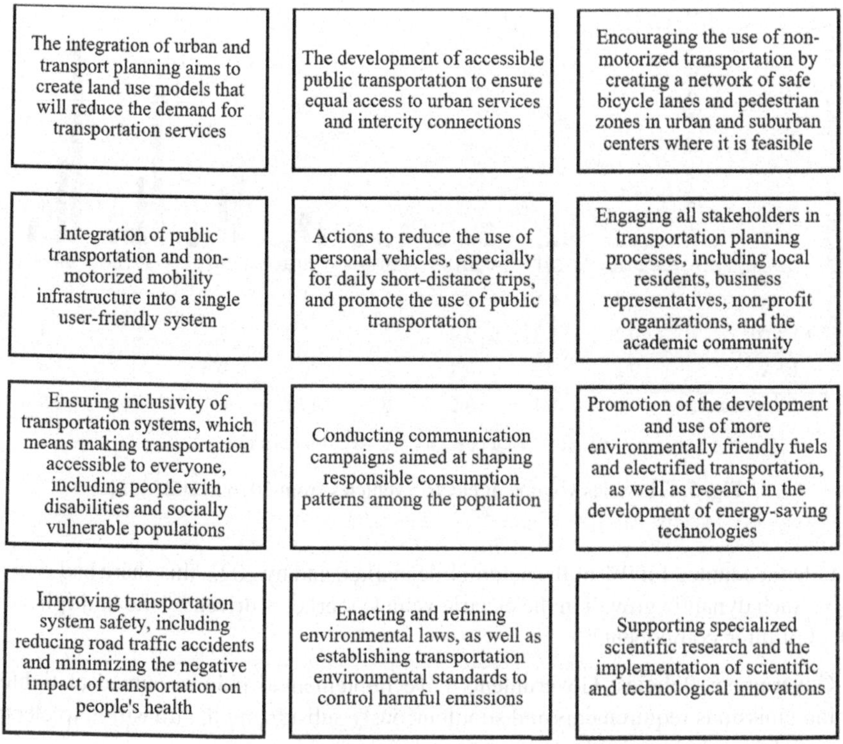

Fig. 4. Proposals from the global expert community for implementing the concept of sustainable mobility in modern cities (compiled by the authors based on [14] materials).

Despite the multifaceted nature of the concept of sustainable mobility, it should be noted that the development and widespread adoption of electric transport play a crucial role and become an indispensable component of urban sustainable mobility.

4 Research Results

In the last decade, the automotive market has experienced a new stage of development, driven by the increasing adoption of electric vehicles in the world's largest economies. Each year, electric cars are becoming more popular among consumers, and manufacturers are intensifying their efforts in the development and production of innovative models. This growing interest in electric vehicles is driven by the rapid advancement of technology and the existing trend of implementing the concept of sustainable mobility in modern cities. Global electric vehicle sales continue to set new records (see Fig. 5). For instance, in 2022, the worldwide electric vehicle sales reached 13.9 million units, despite disruptions in the supply chain, macroeconomic and geopolitical uncertainties, as well as high prices for raw materials and energy resources.

In 2021, the global total sales of electric vehicles exceeded 10 million units, surpassing the overall number of cars sold in the European Union in 2021. In 2021, electric

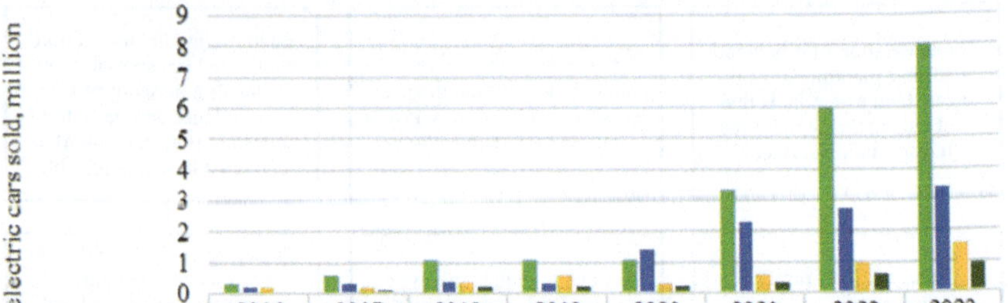

	2016	2017	2018	2019	2020	2021	2022	2023
China	0.3	0.6	1.1	1.1	1.1	3.3	6	8
Europe	0.2	0.3	0.4	0.3	1.4	2.3	2.7	3.4
USA	0.2	0.2	0.4	0.6	0.3	0.6	1	1.6
Other countries	0	0.1	0.2	0.2	0.2	0.3	0.6	0.9

China Europe USA Other countries

Fig. 5. The sales volume of electric vehicles from 2016 to 2023 [15].

vehicles accounted for 9% of the automobile market, and by 2023, this share had risen to 14%. Such dynamic growth in the electric vehicle market is driven by the multiplicative effect of three components:

- Government Policies: Governments have implemented policies aimed at tightening emissions requirements and simultaneously subsidizing the transition to electric vehicles, including the development of charging infrastructure.
- Automakers: Major automotive companies are making significant investments in the development of electrified transport to secure a share of this promising market.
- Consumers: Consumers are becoming increasingly demanding in terms of the environmental and economic aspects of vehicles, with many of them willing to switch to electric cars [16].

Among automakers, models of companies such as Tesla and BYD are leading (see Fig. 6). So, in 2022, the sales volume of the most popular model, the Tesla Model Y, exceeded 770 thousand electric vehicles sold.

The Russian electric vehicle market significantly lags behind global sales statistics. In 2022, the sales volume of electric vehicles amounted to just over 2, 1 thousand units, with the total electric vehicle fleet comprising around 23 thousand units [18]. Despite showing promising growth considering its nascent state, the Russian market faces numerous key challenges that impede the development of the electric vehicle market in Russia. These challenges include the high cost of electric vehicles, an insufficient level of infrastructure development (especially outside major cities, including the lack of "fast charge" charging stations), and a relatively limited variety of models from global automakers available in Russia. Additionally, there is a general increase in vehicle prices or a complete shortage of vehicles in certain classes due to current sanctions and the withdrawal of automakers from the Russian market.

There are various solutions to the current problems described above that are hindering the development of the electric vehicle market in Russia. The high cost of electric vehicles

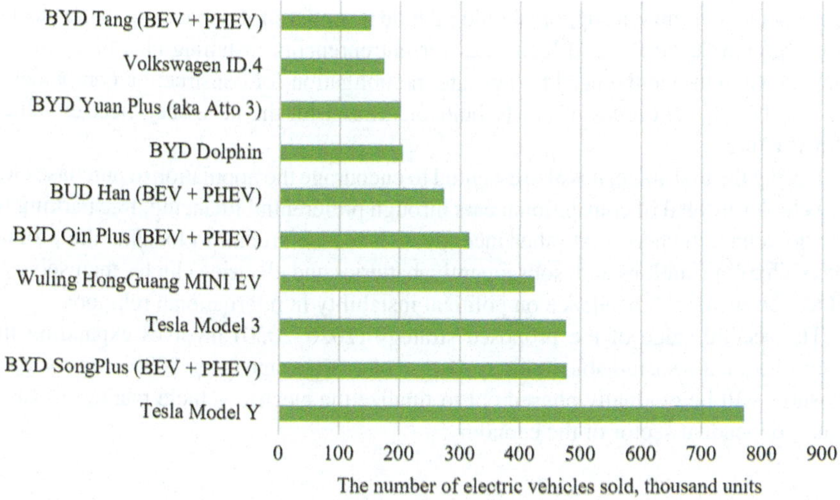

Fig. 6. Sales volumes of the most popular electric vehicle models in 2022 [17]

is attributed to two key factors: a shortage of electric vehicles in the market due to the withdrawal of many global automakers and a high import duty on electric vehicles, which stands at 15%. For comparison, electric vehicles in other Eurasian Union countries are not subject to import duties. The solution to these problems involves seeking new suppliers who are willing to cooperate with the Russian market and reducing the import duty to zero. These measures will help lower the average cost of electric vehicles and meet the demand of consumers. The most promising potential suppliers are from the Chinese market, which is the largest producer of electric vehicles [19]. In China, more than half of the major battery manufacturers for electric vehicles are also represented [20].

Certainly, stimulating the market should involve a comprehensive approach. Thus, the authors of [16] have developed a phased strategy for the development of the electric vehicle market in the Russian Federation. The first stage (until 2025) includes measures to establish a well-developed electric transport infrastructure, particularly charging stations for electric vehicles, and promoting electric vehicles among the population.

One of these measures involves transitioning services that are closely and regularly used by the population, such as taxis, car-sharing, courier services, and food delivery, to electric vehicles. Creating the necessary infrastructure for electric vehicles under this approach would be economically justified due to the extended lifespan of electric vehicles and the profitability of such businesses. It would also generate a positive public perception.

Another potential area where electrification can have a significant positive impact is public transport. In Russia, there is a good example of such electrification in large cities like Moscow, St. Petersburg, Kazan, and others. Since 2017, on public transport routes in these cities, some conventional buses have been replaced with electric buses, and the necessary infrastructure has been developed [16].

However, the establishment of the necessary network of electric vehicle charging stations is not possible without government support. In this regard, the authors of [16] suggest that government support should include incentives for connecting to the electric grid, preferential tariffs, and loan rates for entrepreneurs installing charging stations. They also emphasize the need to legislate the obligation for construction companies to provide the adjacent areas of newly built facilities with the necessary electric vehicle infrastructure.

During the first stage, it is also essential to encourage the population to purchase electric vehicles instead of conventional cars through preferential financing, free parking for electric vehicle owners, and other incentives. It is also crucial to localize the production of charging stations and, subsequently, batteries and electric vehicles themselves to reduce the sector's dependence on political instability in international relations.

The second stage of the proposed strategy (2026–2030) involves expanding this sector into a mass sustainable industry. In the following stage (after 2030), most support measures will be gradually phased out to finalize the electric vehicle market in Russia as an independent sector of the economy.

5 Conclusion

In the modern world, issues related to sustainable development of our society and, in particular, the sustainability of modern cities, are gaining increasing importance. One of the 17 Sustainable Development Goals is "Sustainable Cities and Communities", which outlines the direction for designing modern cities where the rapid development of society coexists harmoniously with the environment, preserving it rather than degrading it. One of the key components of sustainable urban development is sustainable mobility, given the continued growth in the number of cars in cities and the enormous harm they inflict on the environment by releasing hundreds of thousands of cubic meters of pollutants and greenhouse gases into the atmosphere each year, accelerating global warming. This article briefly examined the prerequisites for the development of the concept of sustainable mobility and the level of development of its key component - electric transport. Despite the global trend towards the electrification of transport and high sales of electric vehicles worldwide, the Russian market is only just beginning to gain momentum. To intensify its growth, several problems need to be addressed, such as the high cost of electric vehicles, the low level of infrastructure development, and the limited range of models available.

References

1. Denisenko, M.B., Stepanova, A.V.: Dynamics of Moscow's population over 140 Years. Moscow Univ. Bull. Series 6. Economics. No. 3. (2013). https://cyberleninka.ru/article/n/dinamika-chislennosti-naseleniya-moskvy-za-140-let. Accessed 13 September 2023
2. Environmental Protection in Russia: Statistical Yearbook / Rosstat. – 0-92 M., p. 115 (2022)
3. Gubanova, D., Vinogradova, A., Iordanskii, M., Skorokhod, A.: Variability of near-surface aerosol composition in Moscow in 2020–2021: episodes of extreme air pollution of different genesis. Atmosphere (2022). https://www.researchgate.net/publication/359705741. Accessed 13 September 2023

4. Sassykova, L.R, et al.: The main components of vehicle exhaust gases and their effective catalytic neutralization. Orient. J. Chem. **35**(1) (2019). https://bit.ly/2MZvxjS. Accessed 13 September 2023
5. Wang, B., Wang, B., Lv, B., Wang, R.: Impact of motor vehicle exhaust on the air quality of an urban city. Aerosol Air Qual. Res. **22**, 220213 (2023). https://doi.org/10.4209/aaqr.220213. Accessed 13 September 2023
6. National Center for Biotechnology Information. PubChem Compound Summary for CID 2336, Benzo[a]pyrene. https://pubchem.ncbi.nlm.nih.gov/compound/Benzo_a_pyrene. Accessed 13 September 2023
7. Margolina, I., Veselov, D.: Specific features of noise spreading from motor transport in the urban environment. MATEC Web of Conferences. https://www.researchgate.net/publication/345399891. Accessed 13 September 2023
8. Sustainable Development. United Nations (UN). https://www.un.org/sustainabledevelopment/. Accessed 13 September 2023
9. Costanza, R., Kubiszewski, I.: The authorship structure of ecosystem services as a transdisciplinary field of scholarship. Ecosystem Serv. **1**(1), 16–25 (2012). https://www.sciencedirect.com/science/article/pii/S2212041612000034. Accessed 13 September 2023
10. Agenda for the 21st Century. United Nations (UN). https://sustainabledevelopment.un.org/outcomedocuments/agenda21. Accessed 13 September 2023
11. Sustainable mobility and "smart" connectivity. https://unece.org/sites/default/files/2021-04/2015779_E_web.pdf
12. Transforming Our World: The 2030 Agenda for Sustainable Development / Resolution adopted by the United Nations General Assembly on September 25, 2015 / United Nations (UN). https://www.un.org/en/development/desa/population/migration/generalassembly/docs/globalcompact/A_RES_70_1_E.pdf. Accessed 13 September 2023
13. UNESCO and Sustainable Development Goals / United Nations Educational, Scientific and Cultural Organization (UNESCO). https://en.unesco.org/sustainabledevelopmentgoals. Accessed 13 September 2023
14. Zavyalov, D.: The concept of sustainable mobility in modern megacities. MIR (Modernization. Innovation. Research) (2021). https://www.researchgate.net/publication/355064465. Accessed 13 September 2023
15. IEA, Electric car sales, 2016–2023, IEA, Paris. https://www.iea.org/data-and-statistics/charts/electric-car-sales-2016-2023. Accessed 13 September 2023
16. Semikashev, V.V., Kolpakov, A.Y., Yakovlev, A.A., Rostovsky, Y.K. Development of the electric vehicle market in Russia as a necessary condition for benefiting from the global trend of transport electrification. Ind. Interindustry Complexes **33**, 274–281 (2022). https://link.springer.com/article/10.1134/S1075700722030133. Accessed 13 September 2023
17. Best-selling plug-in electric vehicle models worldwide in 2022 / Statista, Inc. – the global data and business intelligence platform. https://www.statista.com/statistics/960121/sales-of-all-electric-vehicles-worldwide-by-model/. Accessed 14 September 2023
18. -year sales growth 50-fold to 8 200 electric vehicles. https://tadviser.com/index.php/Article:Electric_vehicles_(Russian_market)#.2A_4-year_sales_growth_50-fold_to_8.2C200_electric_vehicles. Accessed 15 September 2023
19. Can you succeed in the china electric vehicles market? Gentlemen Marketing Agency (GMA). https://marketingtochina.com/electric-car-in-china-leader-and-strategy/. Accessed 15 September 2023
20. China has perfectly tangled the battery value chain with electric vehicles - a combo The U.S. and Europe will find hard to beat. Forbes. https://www.forbes.com/sites/jenniferdungs/2023/08/17/china-has-perfectly-tangled-the-battery-value-chain-with-electric-vehicles-leading-to-a-combo-the-us-and-europe-will-find-hard-to-beat/?sh=6f10945e66ab. Accessed 15 September 2023

LNG Terminal Services as a Digital Business Ecosystem

Nikitas Nikitakos[1], Afrokomi-Afroula Stefanako[2], Pavel Kovalishin[3](✉), Elena Gordeeva[4], and Marina Burakovskaya[5]

[1] Sharjah Maritime Academy, Sharjah, United Arab Emirates
[2] Department of Shipping Trade and Transport, University of the Aegean, Chios, Greece
[3] International Relations Department, Baltic Fishing Fleet State Academy of Kaliningrad State Technical University, Kaliningrad, Russia
pavelkovalishinkaliningrad@mail.ru
[4] International Relations Department, Kaliningrad State Technical University, Kaliningrad, Russia
[5] Transportation organization department, Baltic Fishing Fleet State Academy of Kaliningrad State Technical University, Kaliningrad, Russia

Abstract. Today, one of the global market trends in the LNG sector and especially in LNG terminals is the integration of Information and Communication Technologies (ICT) in many areas such as operation, management and efficient planning, commercial and support services, customer contacts, etc. In this direction, many efforts have been made to incorporate various value-added applications and services that implement e-business solutions. This new digital business environment includes the creation and use of innovative software applications and web services by LNG-related companies. Therefore, this upcoming situation requires the conversion of LNG-related companies into a kind of Digital Business Ecosystem (DBE) through the integration of multi-disciplinary approaches based on biology terms and e-business models. The first chapters of the paper discuss the origin of the terms business ecosystems as well as DBE giving a detailed overview. The following chapters describe the DBE framework for LNG terminal services through the development of an innovative business platform capable of hosting and delivering a range of web services and applications. The main findings are summarized in the last section.

Keywords: LNG sector · Information and Communication Technologies · Digital Business Ecosystem

1 Introduction

1.1 A Subsection Sample

Today, the economic, technological, political and social environment in which companies operate is changing rapidly. Many otherwise large companies fail because they could not keep up with rapid technological advances, rendering them obsolete. This situation

R. Polyakov (Ed.): EcoSyst 2023, LNNS 1092, pp. 248–256, 2024.
https://doi.org/10.1007/978-3-031-67354-2_26

needs to be addressed in the context of whole systems, as each firm must be considered as a part of a larger ecosystem or business environment. Although in some cases the use of biology terms may be considered inappropriate or excessive in the business world, it is actually the most appropriate. Representative examples from biology and other physical sciences are often considered an easy and direct way to describe complicated systems concepts. Thus, in the light of the terms of biology the new business environment can be more understandable [1].

Basic principles of this new dynamic business environment are the development of innovative technological solutions and the adoption of Information and Communication Technology (ICT) approaches to business processes. As a result, more traditional companies in various sectors and industries are driven out of the market and replaced by more competitive ICT-based companies [2].

For LNG terminals, as one of the main parts of the whole LNG supply chain that connects LNG resources to end users, the adoption of ICT solutions is of major importance in order to become more competitive and efficient according to the new global market trends. The main LNG station services include import and/or export of Liquefied Natural Gas (LNG), loading and uploading of LNG cargo to/from ocean going tankers, transfer across the site, liquefaction-re-gasification, processing, storage, pumping, compression, metering of LNG, distribution to the consumer market. All the aforementioned areas include ICT approaches such as operation, management and efficient planning, commercial and support services, customer contacts, etc. [3].

Case studies of LNG terminal services around the world based on ICT solutions can provide important insights into the creation of this new type of LNG business ecosystem. This concept is key to improving their trade connectivity and efficiency, providing a wide range of benefits to various stakeholders (e.g. LNG carriers, natural gas system operators, service providers, etc.) and solving a range of issues (e.g. identification and tracking of ships arriving at LNG terminals, loading and unloading rate calculation, delay reduction, warehouse control, storage, and other processes, etc.) [4].

This concept supports the conversion of LNG terminals into Digital Business Ecosystems (DBEs), applying a more interdisciplinary logic based on terms of biology and business models, in order to create an open source business environment that can support continuous evolution and integration of a wide range of web services and apps in the LNG industry.

The first chapters of the paper discuss the origin of the terms business ecosystems as well as DBE giving a detailed overview. The following chapters describe the DBE framework for LNG terminal services through the development of an innovative business platform capable of hosting and delivering a range of web services and applications. The main findings are summarized in the last section.

2 Literature Review-Origin of the Term

The use of the term physical ecosystem as a metaphor to describe the capitalist economy is already used by many sciences, since the capitalist economy can be considered as an alive ecosystem. Natural organisms as well as organisms as a metaphor, are based on complex relationships and hierarchies. In the first case, natural organisms are made up of a number of cells, tissues, and organs, etc. in the wider context of natural populations, while in the second, organisms are made up of a number of individuals, groups, departments, divisions, etc. in the wider context of business environment [5].

James Moore (1993) [6], an American management analyst, first used the term "business ecosystem", including a wide range of sectors and industries. Moore analysed and presented the new business environment in which enterprises compete, coexist and create sustainable growth, replacing the term "industry" with the term "business ecosystem" [7]. A business ecosystem can be seen as an economic group in which enterprises work cooperatively as well as competitively to create new and innovative products and services and to satisfy customer needs. Within a business ecosystem sometimes direct and/or indirect competitors can also transform to business partners as circumstances change. The basic principle arises from the concept of complexity. The importance and contribution of complexity has been widely accepted in various physical sciences, such as biology, and with applications in many fields, e.g. Socio-Economic systems. In the context of business ecosystems, the self-organization, emergence, co-evolution, and adaptation are the main complexity terms applied [8].

The term of Digital Business Ecosystem was first introduced by the Information and Communication Technologies (ICTs) department of the European Commission's DG Information Society and Media in 2002, simply adding the term digital to the Moore's term [9]. So, according to ICT department of the EE, a digital business ecosystem can be depicted as follows:

- Digital (ecosystem): it includes the necessary technical infrastructure (i.e. P2P architecture technology) that can detect, transfer and connect a multitude of services and information through internet connections, enabling a variety of transactions and distribution of digital information available in the infrastructure.
- Business (ecosystem): can be considered as an economic group supported by all interacting organizations and/or people, for example enterprises/organizations of the commercial world.
- Ecosystem: a metaphor which emphasizes the interdependence of all stakeholders involved in the business environment, who together evolve their capabilities and roles respectively.

According to the European Commission (EC) report "Digital Business Ecosystem" [10], the main structural principles, some more general, while others more specific, that should be incorporated in an ecosystem, include the following: No single point of failure or control, b) Digital ecosystems should not rely on any actor involved, c) All actors must have equal access opportunities, d) Expansibility and durability, e) Ability to continuously develop, diversify and self-organize, f) Activation and support self-reinforcing production and processing network, g) Capability to activate global solutions adapted to local or domain specific needs, h) Global solutions that emerge from local and sectoral inputs, and i) Local autonomy.

A digital ecosystem can be considered as a self-organizing digital framework that focuses on creating a digital environment for the organizations and actors involved, improving cooperation opportunities, promoting knowledge sharing, open technologies and business development models (Fig. 1).

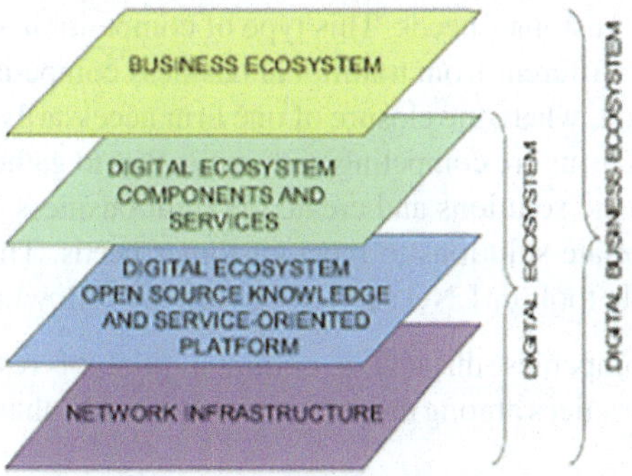

Fig. 1. Description of the concept of Digital Business Ecosystem [11]

Many services are available through Digital Business Ecosystems as users can create, edit and update their profiles, as well as maintain their social contacts (Fig. 2). Users can also initiate a topic network and/or participate to existing networks for discussion and knowledge sharing. Topic authors, technology experts as well as other interested parties can add tagged innovative ideas to a specific topic network, which will be discussed and assessed by the innovation team/group. During monthly in-person meetings, the most interesting ideas will be presented and analysed by the team with aim of selecting, rejecting, or postponing the implementation of an idea. If a candidate idea is selected for implementation, it is then detailed through a multi-layered implementation process [12].

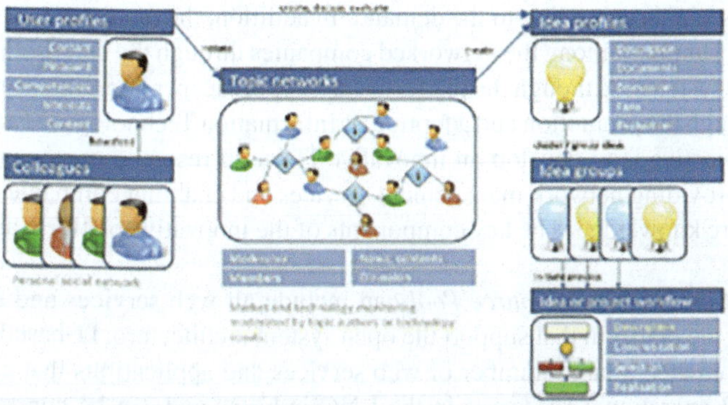

Fig. 2. The Digital Business Ecosystem concept [12]

3 Development of the Innovative Platform

In order to implement the DBE framework to LNG terminal services, an innovative business platform was developed. The platform is based on the Digital Business Ecosystem concept, allowing LNG-related companies to collaborate as well as compete with each other in order to meet customer needs. This type of competition applied in DBE can be characterized as quite different from traditional business competition applied under the rules of the open market, where the closure of one firm necessarily means survival of the others. In this context, firms are competing with each other to gather the necessary workforce, implement holistic solutions and create the right business ecosystem that is able to provide the appropriate solutions to meet customer needs. The innovative platform developed is a valuable tool for LNG services giving the following advantages:

- Networking of companies through a common pool of resources, allowing the exchange of best practices among the parties involved and enhancing the development of LNG market;
- Foment and promote the development of e-business in LNG operations, through the adoption of Information and Communications Technology solutions by LNG companies;
- Institute the framework for LNG companies to act as a group, creating scale economies under the shield of a safe business environment.

3.1 Conceptual Model of the Innovative Platform

The innovative platform developed can be characterized as a digital LNG business ecosystem fully based on network architecture and open source software. Platform can host a number of LNG-related companies (e.g. LNG carriers, natural gas system operators, service providers, etc.) which can benefit from a variety of services available on the platform. Through the platform any LNG-related company can choose a number of resources for e-services and various applications that fulfill its needs. The platform provides a wide range of services as a default option, but all available services are considered scalable according to the demand. In addition, the platform provides collaboration opportunities among the networked companies through the incorporation of their information systems. Although the platform can support deeper integration among stakeholders through the promotion and adoption of Information Technology (IT) approaches, the main objective is to develop an innovative IT-based resource management system capable of providing network management services and enabling companies to collaborate and share knowledge. The key components of the innovative platform are described in Fig. 3.

The first component *Resource Pool* can include all web services and applications available on the platform that support the open system architecture. IT-based technology has already created a large number of web services and applications that can facilitate routines and operational processes in the LNG industry and can be considered as the main resources of the platform. There are also some web services that are valuable tools for most LNG terminals regardless of size and type (e.g. Liquefaction Terminals or/and Regasification Terminals), as they cover a variety of key operational processes for both LNG and maritime sectors (e.g. safety, security, ownership, etc.). These processes require the close cooperation of all parties involved to ensure the smooth commercial operation of an LNG terminal.

The second component *Configuration Models* uses models based on LNG sectors, i.e. LNG subsystems of a Digital Business Ecosystem, including LNG carriers, natural gas system operators, service providers, etc.

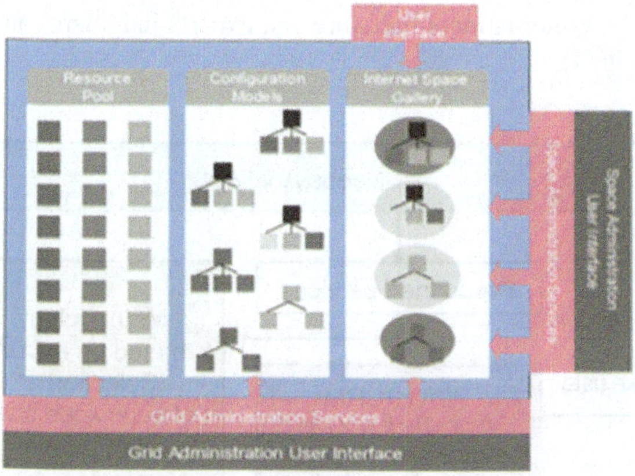

Fig. 3. The key components of innovative platform

The last component *Internet Space Gallery* can be considered as an open portal for every LNG terminal user. Thus, every LNG-related company that is on the platform can benefit from a variety of web services available, aiming to complete its operational processes in the best possible way.

Through server all available applications run properly. Business service providers can be seen as an ideal solution for the transparent execution of web services and business processes. In addition, available software tools can support on-site work and promote cooperation. Although one of the critical issues of operating systems is heterogeneity, the innovative platform is able to host a variety of heterogeneous resources via resource service drivers. Drivers should meet the appropriate criteria and interfaces predefined for each service category.

3.2 Methodology Description

The first stage for the development of the innovative platform is to identify and record the needs within the LNG terminals and the users involved. Thus, a needs analysis is carried out for web services and suitable applications. Needs analysis includes a number of generic requirements and regulations governing LNG terminal services and specific features of LNG markets.

The developed platform should meet the needs for LNG terminal services by offering at all times a wide range of the most wanted web services and rejecting services characterized as low demand. Services characterized as high demand services can take a higher value (score) in the evaluation procedure. Thus, through this process, LNG-related companies of the platform can benefit from useful offers for their operations while at the same time evaluating the quality as well as the quantity of web services available. In any case, objectivity should be ensured during the overall process of evaluating and classifying each available web service based on predetermined evaluation criteria.

The proposed methodology of this paper is based on Multi-Criteria Decision Analysis (MCDA) approach. The set of criteria used to evaluate the web services available on the platform is based on international literature and experts judgments and opinions using questionnaires (Fig. 4).

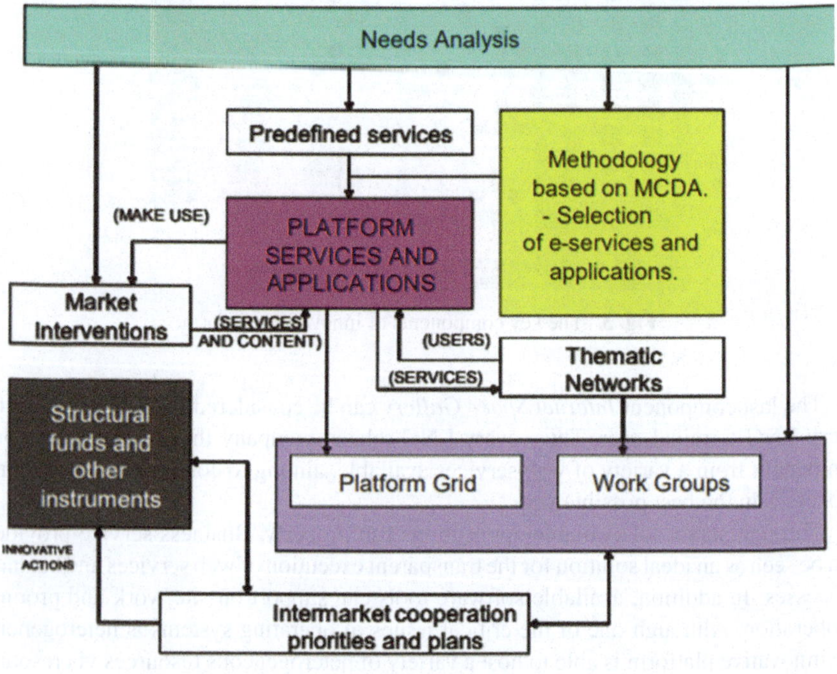

Fig. 4. Methodology flowchart

4 Concussions

The term ecosystem as a metaphor has been widely used to describe the variety of kinds of structures and processes, giving particular emphasis to different aspects of the natural ecosystem and finding applications in a variety of fields. The concepts of the economy as an ecosystem and the social economy are closer to the business ecosystem and therefore can be used as a framework to formulate the concept of business ecosystem. Provided that both of the aforementioned terms are seen as complex, adaptive and constantly evolving systems, it is easier to understand the main rules of their shape, evolution and interdependence in a broader context and to exploit research in other areas of science. Although this issue requires further research, it may also provide important and useful insights to managers in similar business environments.

The LNG sector is a global market with diverse interested parties acting independently of place and time, and this condition essentially links the LNG operation with the creation of networks through the use of ICT approaches.

ICT solutions have already been widely implemented and have modified operational standards in many sectors, including the LNG industry. In the context of ICT approach,

LNG services could be modeled as a Digital Business Ecosystem. Digital Business Ecosystems are a new reality in modern business. The appropriate implementation of DBE in LNG services will facilitate LNG supply chain, by effectively networking the different stakeholders and connecting all available resources into one system, thereby creating an environment capable of sharing business, knowledge and infrastructure. Case studies of LNG terminal services around the world based on ICT solutions can provide important insights into the creation of this new type of LNG ecosystem.

This new type can be implemented through the proposed innovative business platform. The main objective of the platform is to act as an ecosystem using ICT solutions that can increase the overall performance of LNG terminal services without at the same time decreasing the individual performance of companies involved in the digital ecosystem. The conceptual business platform can host and assist a large number of LNG-related companies (e.g. LNG carriers, natural gas system operators, service providers, etc.) by providing a variety of web services and valuable resources that were not available at company or/and sector level and adopting new and innovative ICT approaches to business processes.

Today, scientific research has focused its interest on the concept of Digital Business Ecosystems and a variety of innovative ICT solutions have already been implemented by many industrial sectors. For LNG terminals, as one of the main parts of the whole LNG supply chain that connects LNG resources to end users, the adoption of ICT solutions is of major importance. The rapid development of ICT and its adoption by LNG-related companies can provide a variety of web services in many areas of LNG terminals, including operation, management and efficient planning, commercial and support services, customer contacts, etc. So, this innovative and ever-evolving LNG business environment can be seen as an excellent chance for deeper research in Digital Business Ecosystems.

References

1. Moore, J.: The Death of Competition: Leadership and Strategy in the Age of Business Ecosystems. Harper Paperbacks (1997)
2. Verhoef, P., et al.: Digital transformation: a multidisciplinary reflection and research agenda. J. Bus. Res. **122**, 889–901 (2021)
3. Barateiro, C., Gomez, J.S., Soranz, R.: LNG (Liquefied Natural Gas): the new frontier in metering management. In: 2013 Conference: VII Rio Automation Congress. Rio de Janeiro (2013)
4. Ye, Z., Mo, X., Zhao, L.: MINLP Model for Operational Optimization of LNG Terminals. J. Process **9**(4), 599 (2021)
5. Rothschild, M.: Bioeconomics, Economy as Ecosystem. Henry Holt Publisher (1990)
6. Moore, J.: Predators and Prey: A New Ecology of Competition. Harvard Business Review (1993)
7. Li, W., Vanelslander, T., Liu, W., Xu, X.: Co-evolution of port business ecosystem based on evolutionary game theory. J. Ship. Trad. **5** (2020)
8. Shin, M.M., Jung, S., Rha, J.S.: Study on business ecosystem research trend using network text analysis. Sustainability **13**, 10727 (2021)
9. Nachira, F.: Toward a Network of Digital Business Ecosystems Fostering the Local Development. European Commission, Brussels (2002)

10. Nachira, F., Nicolai, A., Dimi, P., Louarn, M., Leon, L.R.: Digital Business Ecosystems, pp. 4–13. European Commission, Brussels (2008)
11. Chang, V., Uden, L.: Governance for e-learning ecosystems. In 2008 2nd IEEE International Conference on digital ecosystems and technologies, pp. 340–345. IEEE Publishing (2008)
12. Nedbal, D., Brandtner, P., Auinger, A., Erskine, M.A.: The critical mass in collaborative digital business ecosystems for innovation: a case exploration of readiness and willingness. In: 2013 7th IEEE International Conference on Digital Ecosystems and Technologies (DEST), pp. 96–101. IEEE Publishing (2013)

Author Index

R. Polyakov (Ed.): EcoSyst 2023, LNNS 1092, pp. 257–258, 2024.
https://doi.org/10.1007/978-3-031-67354-2